Biophysical Models and Applications in Ecosystem Analysis

Biophysical Models and Applications in Ecosystem Analysis

Jiquan Chen

Higher Education Press
China

Michigan State University Press
East Lansing

⊗ The paper used in this publication meets the minimum requirements of ANSI/NISO Z39.48-1992 (R 1997) (Permanence of Paper).

Michigan State University Press
East Lansing, Michigan 48823-5245

Published in China by the Higher Education Press Limited Company

Published in the United States of America by Michigan State University Press

Library of Congress Cataloging-in-Publication Data is available
ISBN 978-1-61186-393-2 (paperback)
ISBN 978-1-60917-667-9 (PDF)
ISBN 978-1-62895-426-5 (ePub)
ISBN 978-1-62896-427-1 (Kindle)

Visit Michigan State University Press at *www.msupress.org*

Contents

Foreword

As biophysical ecologists we study the metabolism of the biosphere through its breathing. This requires that we observe and model trace gas fluxes between ecosystems and the atmosphere, and the constituent biophysical processes of ecosystems. Modeling involves defining and parameterizing functional relations that describe how trace gas fluxes respond to environmental conditions. Application of these models require that we quantify these functions based on the environmental state that organs (leaves, roots, soil microbes) sense, not those of some distant weather station.

If we want to apply biophysical information for policy, practice and learning, doing so by qualitative means, like waving hands or inferring conditions and relationships by back of the envelope sketches is not good enough. Modern-day scientists, practitioners, policy makers and the next generation of students, need a quantitative toolbox to assess the state of the world in a quantitative manner. Here is where Prof. Chen's book on. *Biophysical Models and Applications in Ecosystem Analysis* can make a mark and fill an educational void.

Students learn better by doing. How to do so effectively and equally, when so many environmental science students have different and wide skill sets? We can provide them with state of art learning from a common base, which will provide hand holding for environmental problem solving. This book meets this goal. It has over 20 examples of biophysical calculations based on spread sheets and Python on a wide range of topics, such as foundational micrometeorological changes, carbon and water fluxes, energy balance, and global warming potentials.

This modeling approach allows students to explore biophysical questions and problems, to ask "how so?" and "so what?" questions and to get immediate feedback from plots. For example, students will be able to explore foundational concepts like the Michaelis-Menten and Arrhenius Equations. They will have in their hands the revolutionary Farquhar's, von Caemmerer, Ball-Berry model of leaf photosynthesis. And, they can study water balances with a suite of evaporation equations.

As someone who started his career modeling with Quick Basic and C, I have learned to realize how blind I was in my early career. In recent years, I have converted many of my codes and scripts to Matlab and Python. In my own teaching I have learned how scripting languages give students an ability to organize their thoughts and how visualization power enables students to inspect the impacts of parameters, different inputs of environmental drivers and predict future scenarios. So, I have first-hand experience of the power and importance of such a suite of well-created and tailored educations tools.

Prof. Chen has produced a much more professional set of applications than those I have cobbled together. Hence, I expect many of my peers will gravitate towards this book for their teaching. And students, will use this book to gain confidence about modeling.

Dennis Baldocchi

Preface

A number of questions usually arise when a student starts to consider quantitative models for fitting their experimental data, achieving a specific study objective, or testing a hypothesis. Common questions include: Which model should I use, and why? Are there alternative options? What are the meanings of the parameters in each model? What values should I use as parameters? And how do I construct the model for my study? My answer to each of these legitimate questions often is, "It depends." This is because every model, biophysically or empirically, was developed for specific conditions or under certain assumptions. In ecosystem studies, no single model could be used to answer the full range of scientific questions. They are based on crucial ecological and physical processes and should not be assumed to work "perfectly" in all ecosystem types, under all kinds of circumstances, and across all spatial and temporal scales. Yet, the answer "It depends" often adds to the confusion, until algorithms are well explained, with history, rationales, and applications. I also find that demonstrative examples with real-world data are very helpful. Over 30 model templates in Excel Spreadsheets or Python codes are provided here for demonstrations and uses. Above needs for context and background were the primary motivation behind writing this book.

A second motivation stemmed from the unprecedented growth in the number and complexity of ecosystem models developed over the past 40 years. Now there are a variety of system models that predict the magnitudes and dynamics of ecosystem properties. Each of these models was carefully constructed with sound algorithms from meteorological, hydrological, ecological, biogeochemical, and/or statistical principles. As a result, they are complex in terms of the number of processes factored, as well as regarding the inter-connections among the processes. Understanding and applying these models are not easy due to their complexity. Fortunately, almost all ecosystem models were developed with a few common algorithms. For example, Farquhar's photosynthesis equation, the Ball-Berry stomatal conductance algorithm, Michaelis-Menten kinetics, temperature-dependent respiration in the form of Q_{10}, and energy balance are widely used. This book is designed to describe and explain the major biophysical and empirical modules that have been used in ecosystem models. Understanding these fundamental algorithms will speed up the application of system models. For model developers, knowledge about each of the crucial modules, including their varieties, behaviors and parameterization, model performances, and their strengths and limitations, are essential to improving and advancing their work. For example, a simple Q_{10} algorithm based on exponential equation (Chapter 3) has been widely used in many ecosystem models for calculating respiration, yet there are many other forms that may provide more realistic predictions, albeit requiring different sets of parameters.

There also are practical reasons to develop a good understanding of fundamental biophysical models (and their algorithms) in experimental studies. Observational and experimental research is conducted with finite amounts of information (*e.g.*, number of manipulative factors). It often is best to express empirical data with quantitative models, so results can be generalized using different combinations of inputs, since observations cannot cover all kinds of environmental conditions. Unexpected results may emerge through this type of modeling exercise. One of the reasons for fitting empirical data into a biophysical model is because it produces meaningful parameters that allow investigators to compare the magnitude, differences and changes of those parameters to help them better understand the underlying mechanisms among ecosystems, over time, or under different environmental conditions (including disturbances). For example, estimated Q_{10} values in soil and ecosystem respiration models have been widely used to assess CO_2 loss from terrestrial ecosystems by types, among continents or climatic zones, and under different natural and anthropogenic disturbances (*e.g.*, wild fires, extreme heat waves). Similarly, V_{max} and J_{max} values estimated in photosynthesis models (Chapter 2) permit us not only to assess the CO_2 assimilations among different leaves, species, or ecosystems, but also to guide management practices such as plantation density, species selection, irrigation scheduling, *etc.* Estimating resource use efficiency as a key parameter in production models is another example for such a purpose.

In some studies, the objective is to re-scale results to different spatial and temporal resolution and content, which requires a mathematically expressed model. Using predicted climate conditions (*e.g.*, from IPCC) to calculate future ecosystem production is an example of this. Recent advancements in remote sensing modeling of ecosystem structure and function also require well-calibrated algorithms to scale up *in-situ* data to regional levels. A well-calibrated biophysical model can serve this purpose. MODIS NPP/GPP products, for example, are based on light-use efficiency (Chapter 2) principles for each time period at a global scale. Another practical use for a biophysical model is to fill data gaps in observations. It is common to have missing data points in observational databases, and filling these data gaps is a necessary step before other analyses can proceed. A great example of this is the continuous measurement of net exchange of carbon, water and energy by using the eddy-covariance method. Due to factors such as stable boundary layers, malfunctions of sensors, climatic conditions, and unexpected disturbances there are many data gaps in these time series. Applying empirical and biophysical algorithms to fill these gaps, or predict other relevant measures (*e.g.*, daytime respiration from nighttime data), are widely used approaches to solve this problem. Again, different protocols and algorithms have been proposed and applied. It is critical to understand the foundation of each algorithm for filling the gaps.

There are many ecosystem functions that can be explored through biophysical modeling. To cover all of them here would not be practical, especially due the cross-disciplinary nature of the processes involved in model construction. The long list of models was necessarily reduced to a few key focal areas while preparing this text. As it stands now, this book covers ecosystem production (Chapter 2), respiration (Chapter 3), evapotranspiration (Chapter 4), and global warming potentials (Chapter 5). Other important processes and functions (*e.g.*, nutrient cycling, species interactions, transpiration, *etc.*) are not covered here, though they are equally important in ecosystem studies. Many of these models were proposed and constructed based on biophysical and biogeochemical processes, suggesting that relevant knowledge bases are needed. Chapter 1 is designed to introduce the major biophysical (mostly micrometeorological) essentials needed to understand the models in the following four chapters. Remote sensing modeling, another emerging field that uses measurements to quantify ecosystem structure and functions, is particularly advantageous in modeling spatial changes (*i.e.* variations within and among ecosystems), but these models are not covered in this book due to rapid advancement in the field.

Many researchers at the Landscape Ecology & Ecosystem Science Lab stimulated ideas and, in some cases, contributed to the content of this book. I am particularly indebted to Michael Abraha, Amy Concilio, Housen Chu, Jared DeForest, Eugenie Euskirchen, Juanjuan Han, James Lemoine, Qinglin Li, Xianglan Li, Cheyenne Lei, Nan Lu, Asko Noormets, Soung Ryu, Pietro Sciusco, Changliang Shao, Ariclenes Silva, Burkhard Wilske, and Terenzio Zenone, for their discussion, data sharing, or insights on some models. Ge Sun has been a long-time collaborator. Recognizing his leading research on water cycle, I twisted his arm to be the lead author of Chapter 4 on evapotranspiration. My appreciation also goes to Huimin Zou, who is in her first year of doctoral studies in Beijing Normal University. She spent hours revising the solar.PY codes and performed the non-linear regression analysis for ecosystem production and respiration models.

As I was finalizing these chapters during the COVID-19 epidemic in the spring of 2020, I was fortunate to have several mentors, colleagues and students provide constructive and detailed reviews and suggestions for improving the drafts. Richard Waring continued his support, which has been steady over the past 30 years, with timely responses, important references, and specific advice regarding the texts on modeling ecosystem production; Martin Kappas helped translate the publications on soil respiration that are in German and held in German libraries; Dan Wang, Mike Abraha and Naishen Liang provided field data on photosynthesis, respiration, and eddy-covariance fluxes for testing models, along with detailed explanations. More than a dozen colleagues provided very detailed reviews, including Devendra Amatya, Altaf Arain, Housen Chu, Jared DeForest, Steve Hamilton, Phil Robertson, Thomas Sharkey, Gordon Smith, Ying-Ping Wang, and Richard Waring. Encouragement from Phil Robertson, Dennis Baldocchi, Jerry Franklin, and Richard Waring to complete this book helped me along. Kristine Blakeslee carefully edited every chapter with sharp eyes and picky wording. Finally, I am grateful for the invitation and persistent support of Bingxiang Li and Yan Guan at the Higher Education Press (HEP). Their trust and confidence in me have been the major force behind the completion of this book. Julie Loehr of MSU Press added her encouragement to promote the Ecosystem Science and Application (ESA) series as a joint endeavor between HEP and MSU Press.

This book would not be possible without support from many foundations and agencies, including the National Science Foundation, NASA (LCLUC, Carbon Cycle Science), USDA NRI, USDA Forest Service, DOE, and State of Missouri. In particular, this book was planned as part of ongoing research at the Great Lakes Bioenergy Research Center (GLBRC) under the U.S. Department of Energy. Multiple resources at GLBRC were accessed to complete the final manuscript.

List of Online Supplementary Materials

Chapter 1
S1-1: Micrometeorological data collected at an open-path eddy-covariance tower located in a switchgrass cropland of the W. K. Kellogg Biological Station (KBS), Michigan, USA, in 2016 (Switchgrass_metdata2016.xlsx). This dataset is provided for modeling exercises (e.g., S1-2, S1-3, S1-4), as well as for model parameterizations in Chapters 2 – 5.
S1-2: A simple simulator of diel change of air temperature based on daily maximum and minimum temperature, as well as their timings (Ta_Diel.xlsx).
S1-3: Calculations of wet-bulb temperature (T_w), dew point temperature (T_d), vapor pressure (e_a), saturation vapor pressure (e_s) and vapor pressure deficit (VPD) from *in-situ* measurements of air temperature (T_a) and relative humidity (h) (Eqs. 1.5 – 1.13) (Ta_h_VPD.xlsx).
S1-4: An empirical model for simulating vertical profile of wind speed over a hypothetical vegetation based on Equation (1.24) (Wind.xlsx).
S1-5: Python codes for simulating the position of the Sun over a given location. Input variables include latitude, longitude, elevation (m), year, month and day of year; outputs are solar zenith angle, declination and day length at 0.1-hour interval (Solar.py).

Chapter 2
S2-1: Light response curves through Michaelis-Menten and Landsberg models (LightResponse.xlsx).
S2-2: Simulations of stomatal conductance (g_s) based on the Ball-Berry model (Ball_Berry_model.xlsx).
S2-3: Field measurements and modeled photosynthesis rate (A_n, μmol m^{-2} s^{-1}) and parameters for two species in Wang *et al.* (2018) (Wang2018.xlsx).
S2-4: Model performances of Michaelis-Menten and Landsberg models for the two species in Wang *et al.* (2018) (LightR_models.xlsx).
S2-5: Python codes for estimating empirical coefficients through nonlinear regression analysis of Michaelis-Menten and Landsberg models (Chap-

ter2_python.rar). This package has one dataset in Excel for practice and four Python programs for non-linear regression.

Chapter 3

S3-1: Spreadsheet models (Schematics.xlsx) for illustrating the roles of two parameters in exponential model (Eq. 3.3) for respiration-temperature relationship, calculations of Q_{10} values, and inclusion of linear constraints by moisture (θ) at high temperature ranges (Fig. 3.2).
S3-2: Field measurements of soil respiration, soil temperature and moisture in 2015 from Chamber #1 (RespirationData.xlsx) in a mature larch plantation (*Larix kaempferi*) (35°26′36.7″ N, 138°45′53.0″ E; 1105 m a.s.l.) on the northeastern slope of Mt. Fuji in central Japan (Fig. 3.3).
S3-3: Spreadsheet modeling and model comparisons (Rmodel_1.xlsx) of linear, exponential and quadratic forms (Eqs. 3.5, 3.3, 3.7).
S3-4: Spreadsheet modeling and model comparisons (Rmodel_2.xlsx) of Logistic, Lloyd-Taylor and Gamma models (Eqs. 3.10, 3.9, 3.11).
S3-5: Spread sheet modeling and model comparisons (Rmodel_3.xlsx) of three model forms by including soil moisture (θ) as an additional independent variable (Eqs. 3.14, 3.16, 3.17).
S3-6: Spreadsheet modeling of soil respiration with day of year (DOY) and soil moisture (θ) as additional covariates of temperature (Rmodel_4.xlsx) (Eq. 3.18) (Fig. 3.7).
S3-7: Python codes for estimating empirical coefficients through nonlinear regression analysis of Logistic, Lloyd-Taylor, Gamma, DeForest, Xu, Concilio and DOY models (Respiration.rar). This file has two Excel data files and 12 Python programs for linear and non-linear regression.

Chapter 4

S4-1: Field measurements of evapotranspiration (ET) and micrometeorological variables at 30 min interval in 2016 in an agricultural site (42°28′36.19″ N, 85°26′48.37″ W, 294 m a.s.l.) with an eddy-covariance tower of the Kellogg Biological Station, Michigan, USA (ETData.xlsx).
S4-2: Spreadsheet modeling of reference ET (ET_o), potential evapotranspiration (PET), and actual ET (Eqs. 4.4, 4.6, 4.12, 4.15, Table 4.4) for a corn field of the Kellogg Biological Station, Michigan, USA (ETModels.xlsx).

Chapter 5

S5-1: Solar radiation intensity and absorptions by different gases through the atmosphere. Data provided by Robert Rhode of Berkeley Earth (AtmosphereTransmission.txt).
S5-2: Spreadsheet model for calculating actual global warming potential

(AGWP) and the relative GWP based on *in-situ* input parameters (GWP_Model.xlsx).

S5-3: Simulations of GWP changes over a 100-year period when a stand is planted for forest from managed cropland (StandDynamics.xlsx).

Scan the QR code or go to
https://msupress.org/supplement/BiophysicalModels
to access the supplementary materials.
User name: biophysical
Password: z4Y@sG3T

List of Symbols

Name	Unit	Full Name
AGB	Mg ha^{-1}; kg m^{-2}	Aboveground Biomass
ANPP	kg ha^{-1} yr^{-1}; g m^{-2} yr^{-1}	Aboveground NPP
aPAR	µmol m^{-2} s^{-1}	Absorbed PAP
BGB	Mg ha^{-1}; kg m^{-2}	Belowground Biomass
BNPP	kg ha^{-1} yr^{-1}; g m^{-2} yr^{-1}	Belowground NPP
C pool	kg ha^{-1}	Carbon Pool
E	mmol m^{-2} s^{-1}; mm	Evaporation
ER	µmol m^{-2} s^{-1}	Ecosystem Respiration
ET	mmol m^{-2} s^{-1}; mm	Evapotranspiration
ET_o	mm	Reference ET
EVI	$(-10000, 10000)$	Enhanced Vegetation Index
fPAR	µmol m^{-2} s^{-1}	Fraction of PAR
G	W m^{-2}	Soil Heat
GPP	Mg ha^{-1} yr^{-1}	Gross Primary Productivity
GWI	—	Global Warming Impact
GWP	—	Global Warming Potential
H	W m^{-2}	Sensible Heat
HFT	W m^{-2}	Heat Flux Transducer
Hs	W m^{-2}	Sensible Heat Flux
L	W m^{-2}	Latent Heat
LAI	m^2 m^{-2}	Leaf Area Index
LE	W m^{-2}	Latent Heat Flux
LSWI	—	Land Surface Water Index
MBC	mg kg^{-1}	Microbial Biomass Carbon
MBN	mg kg^{-1}	Microbial Biomass Nitrogen
NDVI	$(-1, 1)$	Normalized Difference Vegetation Index
NEE	µmol m^{-2} s^{-1}	Net Ecosystem Exchange
NEP	µmol m^{-2} s^{-1}; g m^{-2} yr^{-1}	Net Ecosystem Productivity
NH_4^+-N	mg kg^{-1}	NH_4^+-N
NO_3^--N	mg kg^{-1}	NO_3^--N
NPP	g m^{-2} yr^{-1}	Net Primary Productivity
NUE	—	Nutrient Use Efficiency
PAR	µmol m^{-2} s^{-1}	Photosynthetically Active Radiation
PET	mmol m^{-2} s^{-1}; mm	Potential Evapotranspiration
P_n	µmol m^{-2} s^{-1}	Photosynthesis Rate (A_n)

Symbol	Unit	Description
Q_{10}	—	Temperature Coefficient
R_a	μmol m^{-2} s^{-1}	Autotrophic Respiration
RF	W m^{-2}	Radiation Forcing
R_h	μmol m^{-2} s^{-1}	Heterotrophic Respiration
RH	%	Relative Humidity
R_n	W m^{-2}	Net Radiation
R_{ref}	μmol m^{-2} s^{-1}	Respiration at Reference Temperature
SOC	g kg^{-1}	Soil Organic Carbon
SON	g kg^{-1}	Soil Organic Nitrogen
SR	μmol m^{-2} s^{-1}	Soil Respiration
SWC	%	Soil Water Content
T_r	mmol m^{-2} s^{-1}; mm	Transpiration
T_a	°C; °K	Air Temperature
TH	yr	Time Horizon
TC	g kg^{-1}	Total Carbon
TN	g kg^{-1}	Total Nitrogen
T_s	°C	Soil Temperature
u^*	m s^{-1}	Friction Velocity
VPD	kPa	Vapor Pressure Deficit
VWC	%	Volumetric Water Content
WUE	μmol mmol^{-1}; g kg^{-1}	Water Use Efficiency
α	W m^{-2}	Solar Constant
ψ	degree	Zenith Angle
	MPa	Soil Water Potential
β	degree	Solar Elevation
λ	nm	Wave Length
ϕ	degree	Latitude
δ	degree	Solar Declination
τ	0–1	Sky Transmittance
γ	kPa °C^{-1}	Psychrometric Constant
μ	m s^{-1}	Wind Speed

Chapter 1
Biophysical Essentials for Ecosystem Models

Jiquan Chen

1.1 Introduction

Before getting into the specifics of the models for key ecosystem functions (*e.g.*, evapotranspiration, ecosystem production), it is first necessary to cover some relevant biophysical foundations. Major models for ecosystem production (Chapter 2), respiration (Chapter 3), evapotranspiration (Chapter 4) and global warming potentials (Chapter 5) will be covered in this book. These models, biophysically or empirically derived, have different sets of input variables and/or parameters that are required. Some parameters are physical properties of materials (*e.g.*, vaporization of water at a given pressure), or have been approved as empirically true (*e.g.*, transmission coefficient of light through the atmosphere), while others lack theoretical foundations but can be empirically estimated (*e.g.*, change in atmospheric pressure with elevation). In practice, there also often is mismatch or lack of direct presentation between the required model parameters and the physical quantities that can be measured practically. For example, vapor pressure deficit (VPD) is a critical variable in parameterizing many biophysical models (*e.g.*,

---

Jiquan Chen
Landscape Ecology & Ecosystem Science (LEES) Lab, Department of Geography, Environment, and Spatial Sciences & Center for Global Change and Earth Observations, Michigan State University, East Lansing, MI 48823
Email: jqchen@msu.edu

Jiquan Chen, *Biophysical Models and Applications in Ecosystem Analysis*,
https://doi.org/10.3868/978-7-04-055256-0-1

the Ball-Berry model for estimating stomatal conductance of CO_2 diffusion) (Dewar 2002). It is, however, not directly measured at micrometeorological stations. Hence, it is critical to understand how VPD is calculated by using other measured variables (*i.e.*, air temperature and relative humidity).

A comprehensive introduction to these foundations — the scope of "environmental biophysics" — is a challenge because this book is designed to provide sufficient information to use the models in later chapters. Users are encouraged to consult the classical works of Fritschen and Gay (1979), Rosenberg *et al.* (1983), Kaimal and Finnigan (1994), Griffiths (1994), Chen *et al.* (1999), Lee *et al.* (2004), Ebi *et al.* (2009), Campbell and Norman (2012), Lowry (2013), and Bonan (2019). After an assessment of the major models that will be covered in later chapters, this chapter will cover the following topics:

- modeling diurnal changes of air/soil temperature,
- calculations of atmospheric pressure, VPD, wet-bulb temperature (T_w) and dew point temperature (T_d),
- calculations of zenith and azimuth angles of the Sun,
- calculations of heat flux and/or storage of air, water and vegetation,
- modeling vertical wind profiles of wind speed, and
- quantifications of energy balance equation.

The second challenge that is often frustrating for beginners is the long list of physical (sometimes empirical) constants and the variety of symbols used in environmental biophysics. These constants are described as each occurs and are summarized in the "List of Symbols". Coupled with this challenge is the variety of units that have been used for the same variable. The International System of Units (SI) is applied throughout the book. Calculations for converting between different units are also presented. When possible, symbols and abbreviations are applied consistently throughout the book and are matched with those in Campbell and Norman (2012). Unfortunately, one cannot always avoid using the same symbol for different variables due to the conventions in separate disciplines (*e.g.*, meteorology *vs.* hydrology). For example, R_s is commonly used for "shortwave radiation" in micrometeorology and for "soil respiration" in ecology (*see* Chapter 3).

To best understand the models, their performances and applications, it is effective to use *in-situ* data for demonstrations. Again, this is difficult because of variations in how models perform in different ecosystems, available input parameters, and temporal scale (*e.g.*, hourly-yearly scale) of the applications. While it is desirable to use actual datasets from different ecosystems and for different models, this book is designed as a generic presentation of popular biophysical models. Consequently, I use the data collected from an eddy-covariance (EC) flux tower at a switchgrass (*Panicum virgatum*) bioenergy crop site for demonstrations (S1-1: Switchgrass_metdata2016.xlsx). Other chapters use different sets of data for diverse applications.

The EC tower is one of seven flux towers constructed in November of 2008 by the Great Lakes Bioenergy Research Center (GLBRC) at the W. K. Kellogg Biological Station (KBS), Michigan, USA. It is located within a 14.1 ha plantation of switchgrass (Latitude = 42°28′36.15″N; Longitude = 85°26′48.33″ W; Elevation = 295 m). The region lies on the northeastern edge of the US Corn Belt. The climate is temperate and humid, with a mean annual air temperature of 9.7 °C at KBS and an annual precipitation of 920 mm, evenly distributed throughout the year, with about half falling as snow. The soil textural class of all sites is sandy clay loam with a pH range from 5.8 to 6.4. The site has been cultivated conventionally as a corn/soybean rotation for the past 10 years and was planted in row crops for at least 30 years before that (Zenone *et al.* 2011). In 2009 the field was converted to no-till soybean; it has been planted as switchgrass ever since (Abraha *et al.* 2015).

Continuous open-path eddy covariance and meteorological measurements have been maintained since December 2008. The measurement system includes a LI-7500 open-path infrared gas analyzer (IRGA, LI-COR, Lincoln, Nebraska, USA) for H_2O and CO_2 concentrations and a CSAT3 three-dimensional sonic anemometer (Campbell Scientific Inc. (CSI), Logan, UT, USA) for lateral, longitudinal, and vertical wind velocities and sonic temperature. The sensors are oriented toward the prevailing wind direction to minimize wind distortions due to supporting structures and have been periodically checked and cleaned. The IRGAs is calibrated every 4–6 months using zero-grade nitrogen gas for zeroing H_2O and NOAA standard gas for CO_2 calibration, and a dew point generator (LI-610, LI-COR) and standard CO_2-N_2 gas mixture for setting the H_2O and CO_2 spans, respectively. All EC measurements are conducted at 10 Hz and logged using a Campbell CR5000 datalogger. Additionally, we measure soil heat flux (G) (HFT3, CSI) at 2 cm below the soil surface using three randomly placed soil heat flux plates, soil temperature (T_s) at three depths (2, 5, and 10 cm) below the soil surface using CS107 probes (CSI), and soil water content in the upper 30 cm of the soil profile using a vertically inserted Campbell CS616 time domain reflectometry (TDR) probe (CSI). Measurements of incoming and outgoing short- and long-wave radiation (CNR4 net radiometer, Kipp & Zonen BV, the Netherlands) and air temperature and relative humidity (HMP45C, CSI) are also made at the site. Precipitation is measured at a nearby weather station[1] located about 4 km away from the nearest tower using a tipping bucket rain gauge (TE525WL-S: Texas Electronics, Dallas, Texas, USA). More information about instrumentation and measurements can be found in Zenone *et al.* (2011, 2013) and Abraha *et al.* (2015, 2018).

Data collected in 2016 from this site is used for demonstrations of different models in this chapter. Half-hourly fluxes were computed using the software EdiRe (University of Edinburgh, v1.5.0. 32, 2012) as a covariance of a scalar (sonic temperature, H_2O, or CO_2) and vertical wind speed, following the standard protocols of the FLUXNET (Baldocchi *et al.* 2001, Abraha *et al.* 2015). The datasets have gaps (due to various reasons); they are filled using the gap-filling algorithm of Reichstein *et al.* (2005). Both the raw data and gap-filled data are included for the modeling exercises of this chapter. This dataset is split into several files for convenience of model demonstrations. Altogether, there are 84 variables, labeled with abbreviations and proper units. Table 1.1 lists some of the variables that are frequently used in this book, while the explanations of other variables can be found in the database (S1-1: Switchgrass_metdata2016.xlsx).

1.2 Diurnal Changes of Air Temperature and Humidity

Air temperature (T_a, °C) and relative humidity (h, %) are the two most common variables recorded at climatic stations worldwide and are used in many biophysical models in this book (Chapter 2). They are usually measured at 2.0 m above the ground, but the measurement frequency varies substantially from 30 min to 1 hour, 3 hours, or daily. Prior to automatic weather stations, air temperature and relative humidity were recorded daily for their minimum and maximum values. Modern weather stations are equipped with dataloggers and sensors for continuous measurements a few times per minute, which are tallied at mostly 30 min for the mean, minimum, maximum, range, *etc.* With increasing interest and efforts in modeling ecosystem functions at finer temporal resolution, half-hourly or hourly data are increasingly used in parameterizing biophysical models. This section focuses on a few key issues in modeling and measuring diurnal changes of air temperature and relative humidity.

Air temperature has a unit of Celsius (*i.e.*, centigrade scale) within the

[1] http://lter.kbs.msu.edu/datatables

Table 1.1 Major variables and their units in the KBS-switchgrass dataset (S1-1: Switchgrass_metdata2016.xlsx) for September 22, 2016. Variable names match those in the original databases, with some differences from the symbols applied in this chapter.

Name in the file	Unit	Description
Fc_wpl	mg m^{-2} s^{-1}	Net ecosystem exchange of CO_2, corrected with WPL① method
LE_wpl	W m^{-2}	Latent heat flux density (L)
Hs	W m^{-2}	Sensible heat flux density (H)
tau	kg m^{-1} s^{-2}	Momentum flux (τ)
u_star	m s^{-1}	Friction velocity (u^*)
rho_a_Avg	kg m^{-3}	Moist air density (ρ)
press_Avg	kPa	Atmospheric pressure
wnd_dir_compass	degree	Prevailing wind direction (D)
wnd_spd	m s^{-1}	Average horizontal wind speed (u)
Rad_short_Up_Avg	W m^{-2}	Incoming short-wave radiation
Rad_short_Dn_Avg	W m^{-2}	Outgoing short-wave radiation
Rad_long_Up_Avg	W m^{-2}	Incoming long-wave radiation
Rad_long_Dn_Avg	W m^{-2}	Outgoing long-wave radiation
Rn_short_Avg	W m^{-2}	Average short-wave net radiation
Rn_long_Avg	W m^{-2}	Average long-wave net radiation
Rn_total_Avg	W m^{-2}	Average net radiation (R_n)
t_hmp1_Avg	°C	Average air temperature (T_a)
rh_hmp1_Avg	fraction	Relative humidity (h)
e_Avg	kPa	Average actual vapor pressure (e_a)
VPD_Avg	kPa	Average vapor pressure deficit (VPD)
par_flxdens_Avg	μmol $m^{-2}s^{-1}$	Average flux of photosynthetically active radiation (PAR)
vwc_Avg	%	Average volumetric soil water content of top 30-cm soil
SoilT_Avg (1-3)	°C	Average soil temperature (T_s) at 2 m, 5 m and 10 cm
HFT_Avg (1-3)	W m^{-2}	Average soil heat flux tramsducer (*i.e.*, G_1, G_2 and G_3)

Note: ① WPL, Webb-Pearman-Leuning.

International System of Units. However, in some regions (*e.g.*, the United States, the Bahamas, Belize, the Cayman Islands, Liberia, *etc.*) and in some historical databases it is also given in Fahrenheit (°F). The conversion between the two units is:

$$\text{Temperature (°C)} = 5/9 \cdot (\text{Temperature (°F)} - 32) \tag{1.1}$$

Under standard atmospheric pressure at sea level (~101.3 kPa), the temperature at which water freezes into ice is 32 °F (*i.e.*, zero in °C) and the boiling point of water is 212 °F (*i.e.*, 100 °C). Another unit, the Kelvin scale, is used in biophysics, because many original models, or biophysical relationships, were derived from the absolute zero temperature — the lowest temperature possible — at −273.15 °C.

In a typical mid-latitude ecosystem, air temperature reaches its minimum before sunrise, increases after sunrise, and peaks a few hours after solar noon before decreasing again in the afternoon (Fig. 1.1). Several measures of temperature during the course of a day include daily minimum (T_{min}), maximum (T_{max}), mean (T_{mean}), range (T_{range}), and timing when T_{min} and T_{max} occur. Because of large diurnal changes in air temperature, it is crucial not to use a snapshot measure to describe the temperature of the day. For example, on September 22, 2016, T_{a} at KBS-switchgrass was 24.8 °C at 10:00 h, but it varied from a minimum of 14.3 °C to a maximum of 30.2 °C over the 24 hours.

Air temperature is not always continuously recorded or reported for a study site. In cases where continuously measured air temperature is unavailable, T_{mean} in the literature of popular ecosystem models is often calculated from T_{min} and T_{max} as:

$$T_{\text{mean}} = (T_{\text{min}} + T_{\text{max}})/2 \tag{1.2}$$

Daily mean temperature based on Equation (1.2) is an approximation and can carry large biases. Based on Equation (1.2), the T_{mean} at the KBS site is 22.3 (°C), although it is 21.3 (°C) when calculated from the continuous 30-min time series (S1-2: Ta_Diel.xlsx). An alternative would be to model the diurnal changes before calculating T_{mean}.

Numerous efforts and algorithms have been proposed to model the diurnal changes of air temperature based on T_{min} and T_{max} in the literature (*see* Chen *et al.* 1993b). A model proposed by Parton and Logan (1981) is among the good choices for this purpose owing to its simplicity. Air temperature between sunrise and sunset is calculated as:

$$T_{\text{a}} = (T_{\text{max}} - T_{\text{min}}) \cdot \sin\left(\frac{\pi \cdot m}{D_{\text{L}} + 2\alpha}\right) \tag{1.3}$$

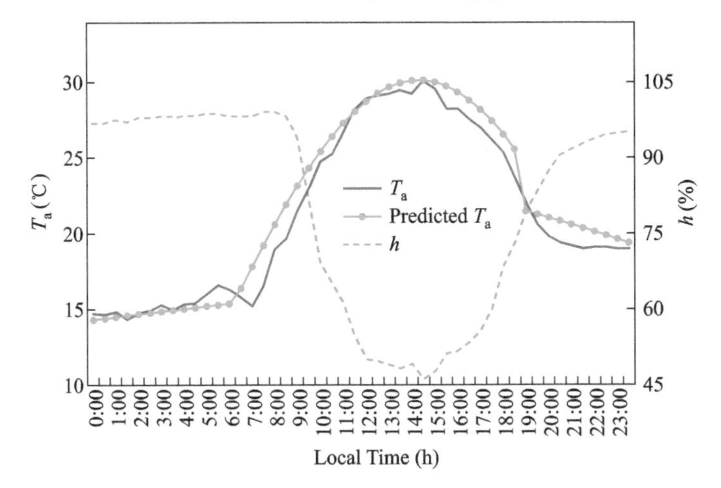

Fig. 1.1 Diel changes of air temperature (T_a) and relative humidity (h) at an experimental switchgrass site (Latitude = 42°28′36.15″N; Longitude = 85°26′48.33″W; Elevation=295 m) of the Great Lakes Bioenergy Research Center (GLBRC), W. K. Kellogg Biological Station (KBS), Michigan (USA) on September 22, 2016. The field data were collected at an eddy-covariance flux tower that will be used throughout this chapter for demonstrations of different biophysical models. Details about tower setup, measurement systems, and data collections and processes can be found in Zenone *et al.* (2013) and Abraha *et al.* (2015). Predicted air temperatures are based on algorithms (Eqs. 1.3 and 1.4) from Parton and Logan (1981), with T_{max} of September 21 and T_{min} of September 23 used for the two sections of nighttime temperature on September 22, 2016 (*i.e.*, before sunrise and after sunset).

where m is the number of hours after T_{min} occurs, D_L is the day length (hour), and α is the lag coefficient for T_{max}. For nighttime after sunset, T_a is calculated as a log linear function:

$$T_a = (T_{min\,0}) - (T_{sr} - T_{min\,0}) \cdot e^{-\beta \cdot \frac{n}{Z}} \tag{1.4}$$

where $T_{min\,0}$ is the minimum temperature of the previous day, T_{sr} is the temperature at sunrise, β is the nighttime temperature coefficient, n is the number of sampling for the day, and Z is the night length (hour). The sunrise/sunset time and length of day/night are dependent on geographic positions (*e.g.*, latitude, longitude, elevation) and day on year (DOY) (*see* Section 1.4). Hourly values of soil temperature can be estimated using these algorithms (Parton and Logan 1981) or the modified algorithms of Kimball and Bellamy (1986). This model assumes an exponential decrease, with lag hours, after sunset and a sinusoidal pattern between sunrise and sunset. For September 22, 2016, at the KBS-switchgrass site, the sunrise and sunset times were 5.79 hour and 17.95 hour, respectively. This model predicts changes at the hourly scale reasonably well (*e.g.*, Fig. 1.1). The daily mean temperature for the given example is 21.8 °C, which is much closer to the actual mean, as it accounts for the asymmetric diurnal pattern of temperature. More importantly, the predicted temperature values can also be used for approximation of the value at given hours when constructing ecosystem models or exploring its empirical relationships with other physical and ecological variables.

The diurnal change of atmospheric humidity, often expressed as relative or absolute values, shows an opposite pattern, although its magnitude is highly dependent on the regional weather conditions (*e.g.*, atmospheric pressure, air temperature) and height of measurements (Kimball and Bellamy 1986, Campbell and Norman 2012). In brief, relative humidity (%) at night is high, decreases rapidly after sunrise, and reaches its minimum at approximately the same time of $T_{\max}$ (Fig. 1.1). Unfortunately, there are no widely accepted models for estimating the hourly values. Nevertheless, users are encouraged to use a combination of linear and sinusoidal algorithms when the daily minimum value and timing for the maximum value as well as for any sudden decreases/increases (often after sunrise/sunset) are known (Chen *et al.* 1993a).

1.3 Atmosphere Water Vapor Pressure and VPD

Water in the atmosphere is crucial for all biophysical processes in terrestrial ecosystems. Amount of water in the air is quantified with water vapor density or vapor pressure. Vapor density is the relative weight of water in the weight of an equal volume of air. Vapor density (or vapor pressure) exponentially increases with air temperature (Fig. 1.2). For a given temperature, the maximum amount of water that a volume of air can hold is called saturation vapor pressure (e_{s}). The difference between actual vapor pressure (e_{a}) and e_{s} is the VPD. In practice, vapor density (pressure) is also reported as relative humidity (h) — the amount of water vapor present in air expressed as a percentage of the amount needed for saturation at the same temperature. With a fixed amount of water in the air, dew forms with temperature decreases — a term called dew point temperature (T_{d}). Under a fixed pressure, the temperature reading from a thermometer covered in water-soaked cloth (wet-bulb) is called wet-bulb temperature (T_{w}), this can be compared with the dry air temperature (T_{a}). At 100% relative humidity, the wet-bulb temperature is equal to the air temperature (dry-bulb temperature). The relationships among T_a, T_d, T_w, e_a, e_s, and h are conventionally illustrated as psychrometric chart (Fig. 1.2). Reading of the chart is based on the above definition.

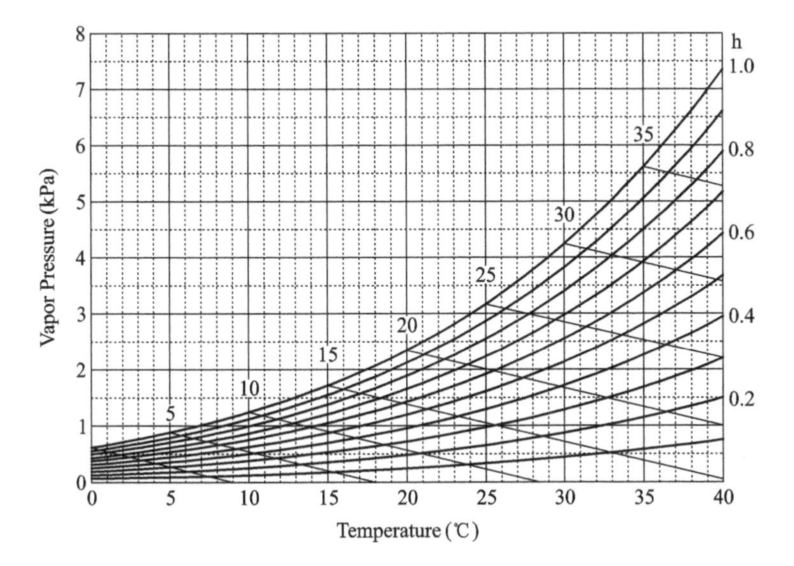

Fig. 1.2 Psychrometric chart of the relationships among ambient air temperature (T_a), relative humidity (h), dew point temperature (T_d), wet-bulb temperature (T_w) and vapor pressure (modified from Campbell and Norman 2012).

Vapor pressure, VPD, and associated biophysical variables are required input variables (*i.e.*, drivers) in many ecosystem models. In practice, however, air temperature and relative humidity are directly measured. Using the 30-min mean air temperature and relative humidity at KBS-switchgrass, this section presents commonly applied algorithms for calculating hourly values of actual vapor pressure (e_a, kPa), actual vapor density (E_a, kg m^{-3}), saturation vapor pressure (e_s, kPa), saturation vapor density (E_s, kg m^{-3}), vapor pressure deficit (kPa), dew point temperature (T_d, °C) and wet-bulb temperature (T_w, °C). These algorithms were well described in detail by Fritschen and Gay (1979), Campbell and Norman (2012), and Lowry (2013).

Atmospheric vapor pressure can be approximated from altitude A (elevation above the sealevel):

$$e_a = 101.3 \cdot e^{-\frac{-A}{8200}} \tag{1.5}$$

where A (m) is the altitude of a site. The vapor pressure at sea level is assumed at 101.3 kPa.

Actual vapor density (kg m^{-3}) is a function of vapor pressure (kPa) and air temperature (°C):

$$E_{\mathrm{a}} = \frac{2170 \cdot e_{\mathrm{a}}}{T_{\mathrm{a}}} \tag{1.6}$$

Saturation vapor pressure can be estimated from Tetens for temperatures above 0 °C (Monteith and Unsworth 2013):

$$e_{\mathrm{s}} = 0.6118\mathrm{e}^{\left(\frac{17.502T_{\mathrm{a}}}{T_{\mathrm{a}}+240.97}\right)} \tag{1.7}$$

Alternatively, Lowry (2013) and Fritschen and Gay (1979) proposed a power function for estimating vapor pressure:

$$\begin{aligned} e_{\mathrm{s}} = {} & (6.1078 + T_{\mathrm{a}}(0.44365185 + T_{\mathrm{a}}(0.01428945 + T_{\mathrm{a}}(0.00026506485 \\ & + T_{\mathrm{a}}(3.0312404 * 10^{-6} + T_{\mathrm{a}}(2.034809 * 10^{-8} \\ & + T_{\mathrm{a}} * 6.1368209 * 10^{-11}))))))/10 \end{aligned} \tag{1.8}$$

Saturation vapor density (E_{s}) can be calculated as:

$$E_{\mathrm{s}} = \frac{e_{\mathrm{s}}}{4.62 \times 10^{-4}(T_{\mathrm{a}} + 273.15)} \tag{1.9}$$

With known relative humidity and E_{s}, vapor density (*i.e.*, absolute atmospheric humidity) is calculated as:

$$E_{\mathrm{a}} = \frac{h \cdot E_{\mathrm{s}}}{100} \tag{1.10}$$

By reversing the Tetens equation (Eq. 1.7), dew point temperature (T_{d}) is estimated as (note that parameters are slightly different) (*see* Fritschen and Gay 1979):

$$T_{\mathrm{d}} = \frac{237.3\log_{10}\left(\frac{e_{\mathrm{a}}}{0.61078}\right)}{17.269 - \log_{10}\left(\frac{e_{\mathrm{a}}}{0.61078}\right)} \tag{1.11}$$

Wet-bulb temperature (T_{w}) is estimated as (Campbell and Norman 2012):

$$T_{\mathrm{w}} = \frac{e_{\mathrm{a}} + \gamma \cdot E_{\mathrm{a}} \cdot T_{\mathrm{a}}}{e_{\mathrm{s}} + \gamma \cdot E_{\mathrm{a}}} \tag{1.12}$$

where γ is the thermodynamic psychrometric constant (0.000666 °C^{-1}), which is calculated as:

$$\gamma = \frac{C_{\mathrm{p}}}{\lambda} \tag{1.13}$$

where C_p is the specific heat of air (29.3 J mol^{-1} °C^{-1}) and λ is the latent heat vaporization of water (40.660 kJ mol^{-1}, or 2.260 kJ kg^{-1}). λ is the amount of energy (enthalpy) that must be added to transform a given quantity of water into a gas (*i.e.*, vapor). It is slightly dependent on air temperature (~0.01% °C^{-1}). Because relative humidity is sometimes more difficult to measure under certain circumstances, direct measurement of wet-bulb temperature, which is relatively easy to measure, has been practiced as an alternative to estimate all other variables using the above algorithms. A word of caution is that these equations should only be applied when air temperature (T_a, °C) is greater than zero. The second caution is that these calculations should not be applied for longer time scales (*e.g.*, more than 2 – 3 hours) because atmospheric vapor density or pressure can be very different. The above equations are nonlinear, suggesting that the mean output from each equation is different from the results when means are used instead. In sum, e_a, e_s, and VPD should be averaged after being calculated based on 30-min or hourly values.

Using the 30-min measurements of T_a (°C) and h (%) at the KBS-switchgrass site on September 22, 2016, a spreadsheet model to calculate other variables of half-hourly values (S1-3: Ta_h_VPD.xlsx) is provided. The diurnal changes in e_s, e_a, E_a, E_s, VPD, T_d, and T_w reflect some typical patterns for major ecosystems in temperate regions (Fig. 1.3). In brief, e_s and VPD change similarly with time, except in the early morning and late afternoon (Fig. 1.3a) when e_a peaks (mostly due to a rapid increase in air temperature). After mid-morning, e_a stabilizes until around 17:00 hours, resulting in a bell-shaped VPD during the day. For wet-bulb temperature, there appears a similar diurnal pattern with relative humidity (Fig. 1.3b and Fig. 1.1). However, the diurnal change in dew point temperature is different, with the twin peaks occurring in early morning and late afternoon (Fig. 1.3b). Nevertheless, these diurnal patterns vary greatly among days and seasons and should not be interpreted as a "typical diurnal pattern" for the site, or elsewhere in temperate regions.

1.4 Solar Radiation

Solar radiation is the ultimate energy source necessary for ecosystems to function. It is conventionally expressed as flux density in units of W m^{-2}, or flux in units of W m^{-2} s^{-1}, depending on one's needs (note that 1 W = 1 J s^{-1}). The amount of energy received at the top of the atmosphere

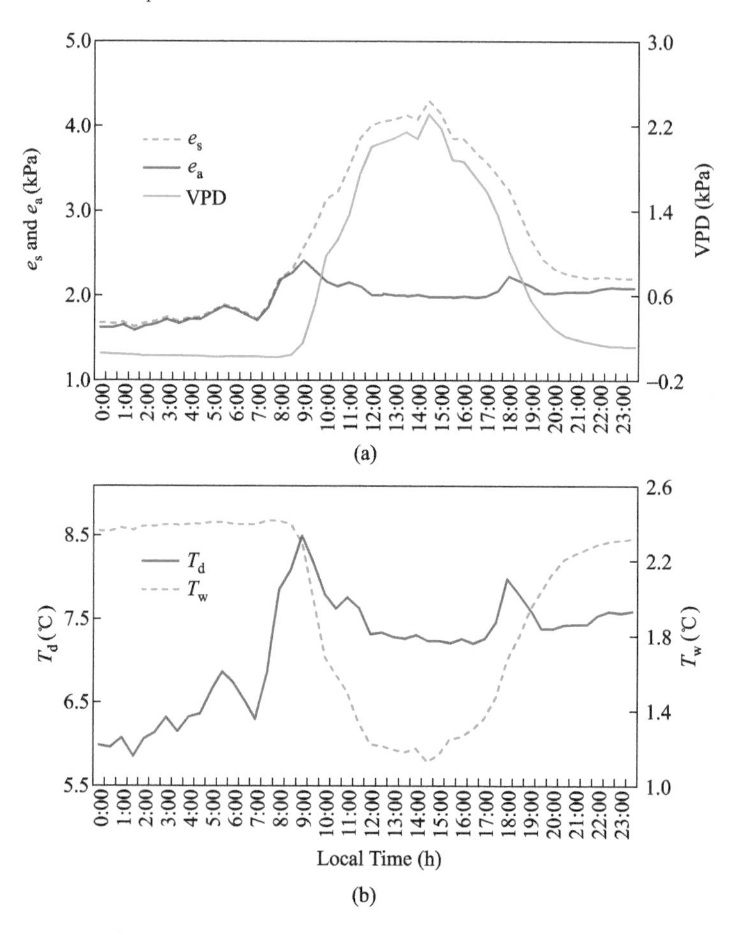

Fig. 1.3 Diurnal changes of key biophysical variables estimated from Equations (1.5) – (1.13) for a switchgrass plantation at the Kellogg Biological Station, MI, USA on September 22, 2016. The model is presented in an Excel spreadsheet (S1-3: Ta_h_VPD.xlsx).

is between 1360 W m^{-2} and 1380 W m^{-2}, commonly known as the solar constant (R_0) (Fig. 1.4).

As sunbeams pass through the atmosphere, a certain amount of incoming solar radiation is absorbed or reflected back to the universe, resulting in a reduction that can be determined by sky transmittance (τ):

$$R = \tau \cdot R_0 \tag{1.14}$$

where τ varies with the path length of solar beams through the atmosphere and air turbidity. Alternatively, R can be modeled with Beer's law:

$$R = R_0 \cdot \mathrm{e}^{-k \cdot z} \tag{1.15}$$

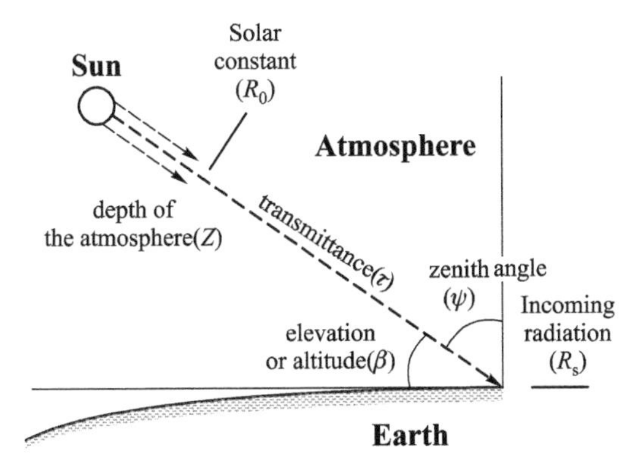

Fig. 1.4 Schematic illustration of key parameters for calculating solar radiation flux density (R_s) at the land surface. Solar constant (R_0) is the radiation flux density normal to the Sun's beams on top of the atmosphere; zenith angle (ψ) is the difference in solar elevation (β) from 90 degrees (*i.e.*, $\psi = 90 - \beta$); solar flux density normal to the Sun's beam (R) is determined from R_0 and sky transmittance (τ, Eq. 1.14), or a combination of atmospheric extinction coefficient (k, km^{-1}) and the path length of solar beams through the atmosphere (z, km) (Eq. 1.15).

where k is the atmospheric extinction coefficient (km^{-1}) and z (km) is the path length through the atmosphere, which depends on the solar elevation (β, degree) and solar declination (τ). Rosenberg *et al.* (1983) reported k values from 0.01 km^{-1} in very clear air to 0.03 – 0.05 km^{-1} in turbid air. Clouds, aerosols, and other particulate matters in the air directly determine the k value. As expected, the z value changes rapidly with hour of a day and is larger in the early morning or late afternoon than at noon. Nevertheless, solar radiation at the land surface normal to Sun's beams should be always lower than the solar constant. The horizontal flux density of solar radiation at the land surface is calculated with cosine law (Fig. 1.4):

$$R_s = R \cdot \cos\ (\psi) \tag{1.16}$$

where ψ is zenith angle (degree) ($\psi = 90 - \beta$). At the KBS-switchgrass site, R_s in 2016 ranged from 450 – 500 W m^{-2} in December and to 900 – 1000 W m^{-2} in June (S1-1: Switchgrass_metdata2016.xlsx).

Atmospheric transmission varies by wavelength (*e.g.*, infrared, visible), latitude, time of day, day of year, altitude and atmospheric conditions (*e.g.*, temperature, pressure, humidity, chemistry). The essential parameters for calculating the R_s value of a specific location (Eq. 1.16) include sunrise time, sunset time, solar declination, solar elevation or zenith angle, and atmospheric transmittance, where sunrise and sunset define the beginning and end boundaries of R_s. The following widely used algorithms in environmental biophysics can be used to approximate these values (*see* Campbell and Norman 2012).

Sunrise time (Time_{sr}, hour) is calculated as:

$$\text{Time}_{\text{sr}} = 12 - \frac{1}{15}\cos^{-1}\left(\frac{-\sin\phi\sin\delta}{\cos\phi\cos\delta}\right) - \frac{t}{60} \tag{1.17}$$

where t is local time (hour). Sunset time (Time_{ss}) is calculated as:

$$\text{Time}_{\text{ss}} = 12 + \frac{1}{15}\cos^{-1}\left(\frac{-\sin\phi\sin\delta}{\cos\phi\cos\delta}\right) - \frac{t}{60} \tag{1.18}$$

Zenith angle (ψ) is calculated from:

$$\psi = \cos^{-1}\{\sin\phi\sin\delta + \cos\phi\cos\delta\cos\ [15(t - t_0)]\} \tag{1.19}$$

where t_0 is the time of solar noon. Solar declination (δ) from 23.45° (*i.e.*, summer solstice) to −23.45° (*i.e.*, winter solstice) is calculated as:

$$\delta = -23.25 \cdot \cos\left[\frac{360}{365}(d + 10)\right] \tag{1.20}$$

where d is the day of the year (DOY), with January 1 as 1. Occasionally, biophysical models may also need the solar azimuth angle (A_z), which is the Sun's relative direction along the local horizon. It is usually applied the same as compass directions, with North = 0° and South = 180°. At solar noon, the Sun is always directly south in the northern hemisphere and directly north in the southern hemisphere. A_z can be calculated as:

$$A_z = \cos^{-1}\left(\frac{-\sin\delta - \cos\delta\sin\psi\sin\phi}{\cos\phi\sin\psi}\right) \tag{1.21}$$

For more advanced calculations, with updated coefficients, users can consult the publications of the Nautical Almanac Office of the United States Naval Observatory[1]. Based on the "Nautical Almanac of 1989", a Python-based calculator was developed for the users of this book to visually assess the changes in sunrise time, sunset time, day length, solar zenith angle, and solar azimuth angle (degree) (S1-5: Solar.py). This model requires inputs of year, month of the year (1–12), day of the month (1–31), latitude (degree with decimals), longitude with decimals, and time interval for output. Sunrise time, sunset time, zenith angle and azimuth angles are calculated with a graphical display. Many similar online calculators are available (*e.g.*, the Earth System Research Lab of NOAA[2]. To demonstrate its use, the diurnal

1 https://bookstore.gpo.gov/agency/nautical-almanac-office

2 https://www.esrl.noaa.gov/gmd/grad/solcalc/sunrise.html

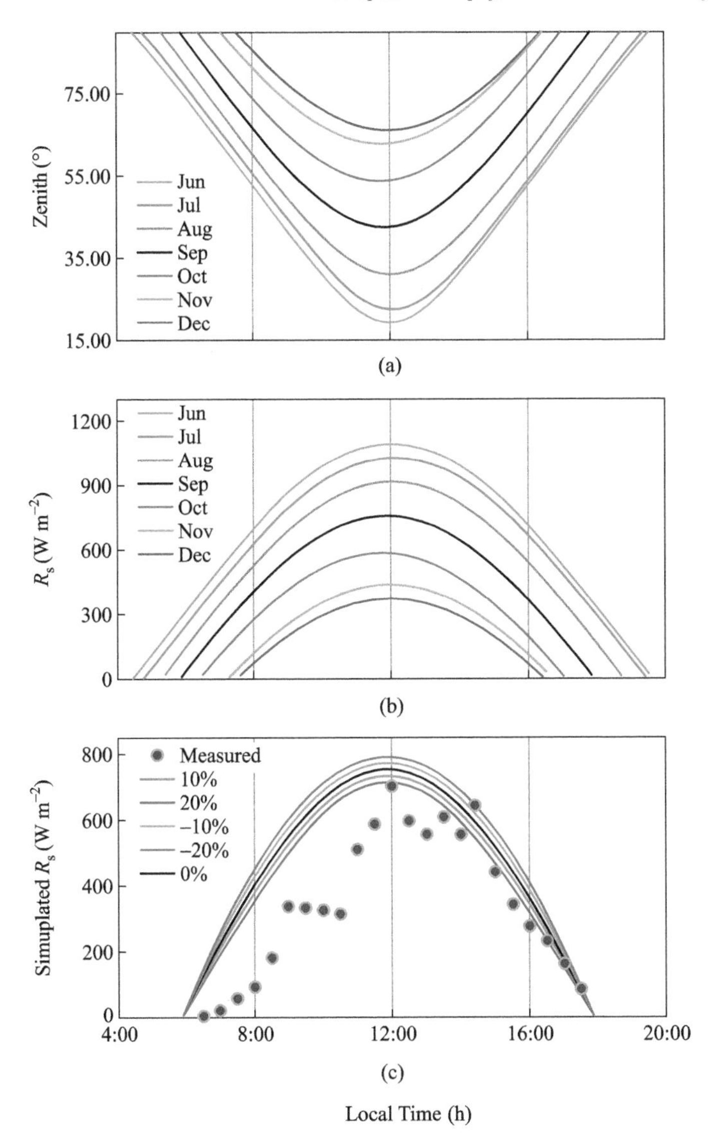

Fig. 1.5 (a) Diurnal change of solar zenith angle (degree) for the 22nd day of June–December in 2016 at the KBS-switchgrass site; (b) simulated incoming solar radiation (R_s, W m^{-2}) by assuming a transmittance of 0.85 on June 22 and a monthly decreasing rate of 4% for the seven days during June–December based on Equations (1.17) – (1.19); and (c) the simulated/measured R_s values for September 22, 2016, by assuming ±10% and ±20% variation of sky transmittance from the mean value (*i.e.*, 0.75) used in (b).

changes in zenith angle on the 22nd day for June through December of 2016 are presented in Fig. 1.5a for the KBS-switchgrass site. The sunrise and sunset times for June 22, 2016, were 4.5 hour and 19.5 hour, respectively, which yielded a day length of 15 hours; for December 22, these values are 7.5 hour and 16.5 hour, respectively, and resulted in a day length of 9 hours. More importantly, the minimum zenith angle at local noon was 19.04 degrees on June 22, but was 61.95 degrees on December 22, which determines the maximum and diurnal changes of radiation flux density on the ground (compared with sky transmittance).

Atmospheric transmittance value (τ) is another key variable in calculating the actual radiation flux density at the land surface (Fig. 1.4). While tremendous efforts have been made in the literature to quantify the mechanistic influences of atmospheric characteristics (*e.g.*, chemical and physical properties of the air), a fundamental knowledge base is the understanding of spatial (different latitude and longitude) and temporal (diurnal, seasonal) changes. A simple Excel spreadsheet model for this purpose is provided to demonstrate these changes by using the KBS-switchgrass site. Assuming $\tau = 0.85$ for June 22, 2016, the peak R_s value would be 1086 W m^{-2} (Fig. 1.5b). With a monthly decreasing rate of 4%, τ would be dropped to 0.66 (*i.e.*, 66% of radiation reached the ground) and result in an R_s value of 430 W m^{-2} at noon. This reduction is partially due to the decrease in zenith angle in December (Fig. 1.5b). These simple simulations are for illustration purposes; alternative approximations in the literature include using a τ value of 0.852 or processing remote sensing images when *in-situ* measurements are unknown[1]. Nevertheless, users can use different τ values to see how they may directly affect R_s for their study sites.

Atmospheric transmittance value also changes substantially over the course of a day because of coupled changes in azimuth angle, diurnal change in air turbidity, and solar elevations that directly determine the extinction coefficient and the atmospheric depth (Eq. 1.15). One can assume different levels of τ from its hypothetical daily mean to assess its influence on R_s. Here, for September 22, 2016, the illustration model considered a sinusoidal change (based on instantaneous zenith angle) during daytime (*i.e.*, the lowest at noon with decreases following a sinusoidal trend toward sunrise/sunset time) and ±10% and ±20% deviation from the mean. The model predicts the diurnal change of R_s under a clear sky condition, which appears consistent with measured values at the KBS-switchgrass site (Fig. 1.5c), especially for the afternoon (note that it was cloudy before noon). Again, these demonstrations are designed for exploring the potential effects of τ on R_s and for understanding the potential magnitude and diurnal changes for a site. However, this model can be potentially used to answer some specific scientific

1 *e.g.*, https://atmcorr.gsfc.nasa.gov/

questions. For example, the differences between predicted and modeled R_s values would indicate the sky condition (*i.e.*, low transmittance, or cloud cover/depth). Atmospheric transmittance is one of the key parameters in calculating radiation forcing (RF) in modeling global warming potentials (Chapter 5).

Two relevant measures of radiation in ecosystem studies are albedo (α, %) and photosynthetically active radiation (PAR, μmol m^{-2} s^{-1}). Albedo is the amount of reflection of solar radiation out of the total incoming solar radiation. By definition, albedo calculation should include both shortwave (loosely defined for wavelengths of 300–700 nm) and longwave radiation (>700 nm). However, outgoing longwave radiation is more greatly influenced by the Earth's temperature according to Stefan-Boltzmann's Law (*i.e.*, disproportional to the longwave radiation from the Sun) (Chapter 5). As a result, albedo is often calculated based on shortwave radiation. Albedo values can vary from 14%–26% in grasslands, 30%–40% in dry deserts, and <10% in coniferous forests. Albedo also varies by time, latitude, altitude and topography. At the KBS-switchgrass site, the albedo value in 2016 was generally lower than 20%, with slightly higher values in the winter months.

PAR in wavelengths of 400–700 nm, which is close to visible light (370–600 nm), has been especially important in modeling many ecosystem processes because this is the energy used for photosynthesis (Chapter 2). The unit for PAR is often given in μmol photon per m^{-2} s^{-1}. Some authors have attempted to convert this to regular energy flux density units (*i.e.*, W m^{-2}) of short or total radiation, but with large uncertainties due to *in-situ* atmospheric and vegetation conditions. This is because of a different energy distribution of the Sun and different absorptions of solar radiation by the atmosphere. A substantial portion of the bias results from two factors: (1) PAR only covers a portion of the spectra of shortwave radiation, and (2) the conversion of quantum and energy flux is wavelength dependent (Chapter 5). Nevertheless, the diurnal change pattern of PAR is similar to that of R_s (Fig. 1.5c) but with a slightly different shape. Absorption of light by chlorophyll takes place largely within narrow bands that peak at 680–700 nm. Three types of PAR values are frequently calculated in the literature for modeling photosynthesis or gross primary production of vegetation. Intercepted PAR (iPAR) is the amount of PAR caught by various canopy layers as the PAR incident at the top of the canopy travels down through the canopy layers to the ground. Absorbed PAR (aPAR) is the amount of PAR absorbed by canopy leaves. fPAR is the fraction of the incident PAR that is either intercepted or absorbed. Obviously, all three measures are directly affected by vegetation structure (*e.g.*, canopy cover, canopy height, leaf area and vertical distribution, sky condition, *etc.*). aPAR in particular has been a focal variable for input in many photosynthesis-based models of water and carbon (Chapters 2 and 4).

1.5 Heat Storages in Soil, Air and Vegetation

The amount of heat energy stored in soil, air and vegetation is a key measure for explaining the changes in microclimate (*e.g.*, temperature, moisture) and/or energy balance of an ecosystem (*see* Section 1.7). In developing biophysical models for ecosystem functions such as evapotranspiration (ET) and energy balance (Chapter 4), one needs a reliable estimation of heat stored or passed through layers of vegetation, air within the canopies, and the soils (Lindroth *et al.* 2010). This energy term is difficult to measure directly, but is estimated based on the thermal and physical properties of air, organic materials and soil. The following two basic models use a topsoil layer as an example but this can also be applied for air and vegetation (*see* example in Oliphant *et al.* 2004).

In principle, heat passes through a thin layer (*e.g.*, surface soil layer) because the temperatures on the top [$T_s(0)$] and the bottom [$T_s(1)$] are different (ΔT_s,° C). The heat flux density (G, W m^{-2}) is calculated as (Fig. 1.6):

$$G = \kappa \cdot \frac{\Delta T}{d} \tag{1.22}$$

where κ (W m^{-1} K^{-1}) is the thermal conductivity of the soil and d (m) is the thickness of the soil layer. The κ value is $\sim$ 2.5 W m^{-1} K^{-1} for soil minerals, $\sim$1.92 W m^{-1} K^{-1} for organic matter, and 4.18 W m^{-1} K^{-1} for water. Following Equation (1.22), G can be quantified by measuring the temperature difference if the thermal conductivity of the media is known. In ecosystem studies, the κ value of the air, vegetation, and soil is also largely determined by moisture in the media. Fortunately, both temperatures and G can be directly measured with modern sensors, suggesting that modeling the continuous change of κ is possible by reversing Equation (1.22).

As heat passes through the soil, a certain amount is stored in the soil layer. The heat storage (ΔS, W m^{-3}) over a period of time (t) can be calculated as:

$$\Delta S = (\rho_b \cdot c_d + \theta \cdot \rho_w \cdot c_w)\frac{\Delta T}{\Delta t} \cdot d \tag{1.23}$$

where ρ_b (kg m^{-1}) is the soil bulk density, ρ_w is the density of water, c_d

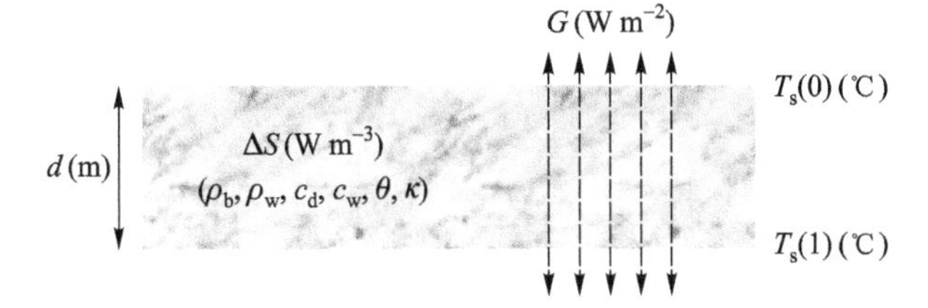

Fig. 1.6 Schematic illustration of heat flow through, and storage in, a thin plate (*e.g.*, topsoil layer), labeled as G and ΔS, respectively. Temperature difference on the two sides of the plate and soil properties are jointly determined the magnitude and dynamics of G and ΔS, including thermal conductivity (κ), density, water content (θ) and specific heat capacity of the soil.

(890 J kg^{-1} K^{-1}) and c_{w} (4190 J kg^{-1} K^{-1}) are the specific heat capacities of the dry mineral soil and the soil water, respectively, θ is the volumetric soil water content (%), and $\Delta T/\Delta t$ (K s^{-1}) is the mean soil temperature change at the time t interval (*i.e.*, can be approximated with the mean values of $T_{\mathrm{s}}(0)$ and $T_{\mathrm{s}}(1)$ at a given time). In terrestrial ecosystems, heat flux density through the soil surface ranges from several W m^{-2} to 10s W m^{-2} in forests and croplands, but it can be high (>100 W m^{-2}) in drylands where canopy cover is low (Oliphant *et al.* 2004, Shao *et al.* 2017). ΔS is small within the canopy column because of the low specific heat capacity of air (29.3 J mol^{-1} C^{-1} *vs.* 75.4 J mol^{-1} C^{-1} for water), while it is also low in vegetation due to the small volume.

1.6 Vertical Profile of Wind Speed

Energy and materials enter and leave ecosystems through the boundaries between vegetation and the atmosphere in gas, liquid and solid forms, assuming horizontal input and output are equal. Other than water, which can be in the form of a liquid (rain, dew) or solid (snow, hail), gas is the dominant form for the exchange of CO_2, N_2O, CH_4, and water vapor. For this reason, it is critical to understand how air moves between the boundaries — a scientific field known as boundary layer meteorology (*see* details in Kaimal and Finnigan 1994).

The wind profile of homogeneous vegetation under neutral atmospheric conditions is typically expressed as a logarithmic function of height (z) from the ground, where horizontal wind speed (U) decreases at heights approaching the ground, as a consequence of surface's drag effects (Fig. 1.7):

$$U(z) = \frac{u^*}{\kappa} \cdot \ln\left(\frac{z-d}{z_0}\right) \tag{1.24}$$

where $U(z)$ is the horizontal wind speed (m s^{-1}) at height z (m); u^* is the friction velocity (m s^{-1}), κ is the von Karmon constant with an average value of 0.35 – 0.43 (Kaimal and Finnigan 1994) (note: a value of 0.40 is often used in the literature), d (m) is a zero plain displacement, and z_0 is the roughness length (m) at which is U is near zero. Because the surface of vegetation is not solid, wind speed at zero depth is not zero; instead it reaches zero at some depth (d) within the top canopy: $U(z_0 + d) = 0$.

u^* depends on the shear stress (τ, kg m^{-1} s^{-2}) at the boundary of the flow and air density (ρ):

$$u^* = \left(\frac{\tau}{\rho}\right)^{\frac{1}{2}} \tag{1.25}$$

where ρ depends on air temperature and pressure. With atmospheric pressure of 101 kPa, the value of ρ at 20 °C is approximately 41.6 mol m^{-3}. u^* represents a characteristic velocity of the airflow and hence reflects the effectiveness of turbulent exchange at the boundaries between vegetation and the atmosphere. Because u^* as a scalar of the model is physically, or empirically, related to many turbulent properties and reflects the canopy structure of vegetation, it has been widely used in constructing biophysical models and has been used as an important indicator for exploring the underlying mechanisms in the literature. For example, u^* has been widely used as an indicator of air mixing in eddy-covariance studies through identifying thresholds of u^* for filtering out unreliable flux measurements below which turbulence may be poorly developed and the results from eddy-covariance method may be biased (Baldocchi *et al.* 2001, Papale *et al.* 2006). By solving the wind profile equation with U and u^* (which commonly can be directly measured by eddy-covariance measurement systems), one can infer the change in canopy height from the changes in z_0 and d (Pennypacker and Baldocchi 2016, Chu *et al.* 2018).

The logarithmic wind profile model (Eq. 1.24) is strictly valid only for the neutral atmosphere (*i.e.*, neither stable nor unstable conditions) and should not be applied for tall canopies or vegetation with large variation across the horizontal space. Other modified versions of the model have been proposed for different purposes (*e.g.*, Goulden *et al.* 1996, Kaimal and Finnigan 1994, Campbell and Norman 2012, Chu *et al.* 2018). Many biophysical models use aerodynamic conductance – a key parameter controlling the exchange of energy/material between vegetation canopies and the atmosphere. For example, u^* has been applied as (Gu *et al.* 2005, Shao *et al.* 2008, Cleverly *et al.* 2015):

$$g_a = \frac{1}{\dfrac{u}{u^{*2}} - 6.2u^{*-0.67}} \tag{1.26}$$

where g_a is used for estimating canopy surface conductance (g_c, m s^{-1}) in the Penman-Monteith equation (Monteith and Unsworth 2013) for estimating latent heat (L) or evapotranspiration (ET) of an ecosystem (*see* Chapter 4). An Excel spreadsheet simulation model is presented in S1-4 (Wind.xlsx) to show how each of the variables in Equation (1.24) may affect the changes of other variables (*see* examples in Fig. 1.7).

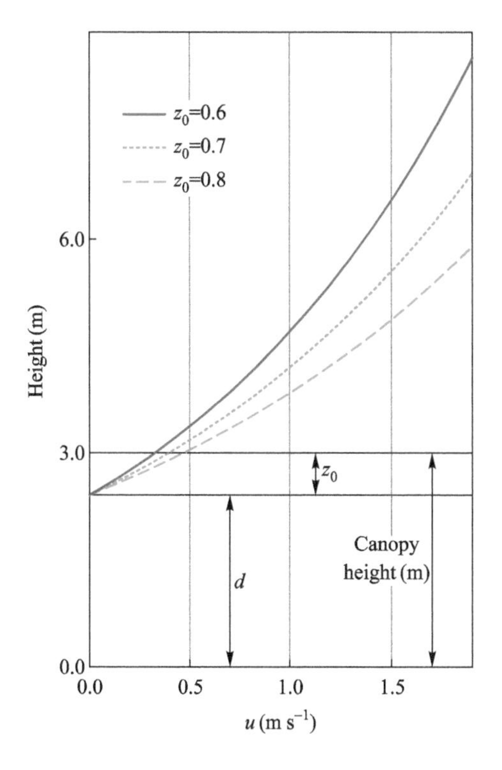

Fig. 1.7 Typical wind profiles over terrestrial ecosystems can be expressed with a logarithmic function (Eq. 1.24). These profiles are simulated by assuming a vegetation height of 3 m (*e.g.*, the peak height of KBS-switchgrass site), a u^* of 0.1 m s^{-1}, κ value of 0.4, and z_0 values of $(z - d)$, with three d values of 2.4 m, 2.3 m and 2.2 m for the illustration of model behaviors (S1-4: Wind.xlsx).

1.7 Energy Balance

Assuming a homogeneous vegetation, the energy balance of a terrestrial ecosystem is conventionally described as:

$$R_n = H + L + G + \Delta S + \varepsilon \tag{1.27}$$

where R_n is net radiation (*i.e.*, incoming – outgoing radiation), H is the sensible heat, L is the latent heat through vaporization (*i.e.*, evapotranspiration, ET), G is the soil heat flux, ΔS is the heat storage over a period of time within the canopy column (air and vegetation), and ε is the energy used for photosynthesis (which is very minor and negligible). $(L + H)$ is commonly called available energy. All terms have a unit of W m^{-2}. ET is also widely expressed with a unit of mm in order to compare with precipitation (mm); a conversion is made by using latent heat of vaporization of water (λ, J mol^{-1}). The value of λ varies by temperature and air pressure. At air pressure of 101 kPa, its value is 44.6 kJ mol^{-1} at 10 °C and 44.1 kJ mol^{-1} at 20 °C (note the molecular mass of H_2O is 18 g mol^{-1}).

While this model has been widely used at multiple spatial and temporal scales, it is more valid for a homogeneous stand and at relative short temporal resolution (*e.g.*, less than a few hours). It is often and appropriately applied to describe the diurnal changes of an individual term or all of the terms (Fig. 1.8a). In theory, energy balance of an ecosystem should hold at any spatial scale, but it is difficult to measure all the terms accurately at the same temporal and spatial scale, with storage terms in the soil and vegetation being especially difficult to measure. The footprint of the corresponding sensors also varies substantially, often resulting in a large portion of missing energy (Fig. 1.8b).

In the literature, the ratio between H and L is called the Bowen ratio (β). This ratio was originally proposed as an indirect method to estimate L and H based on the vertical gradient of temperatures when both L and H are difficult to measure. Using the Bowen ratio-Energy Balance Method, L can be estimated as:

$$L = \frac{R_n - G}{1 + \gamma \cdot \dfrac{\Delta T}{\Delta e}} \tag{1.28}$$

where β can be derived expressed as:

$$\beta = \gamma \cdot \frac{\Delta T}{\Delta e} \tag{1.29}$$

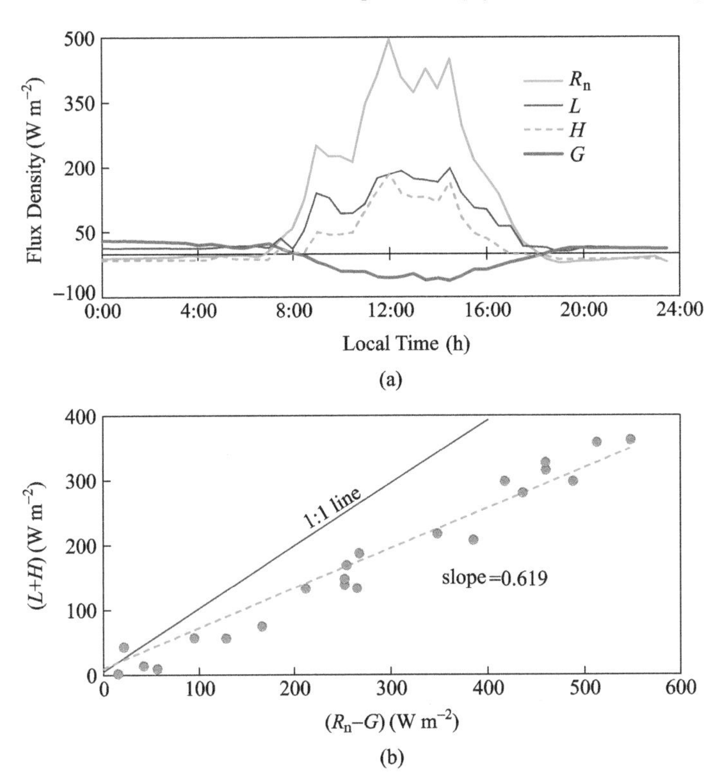

Fig. 1.8 (a) Diel changes of four major energy flux terms (Eq. 1.27) at KBS-switchgrass on September 22, 2016; (b) relationship between $(L+H)$ and (R_n-G) showing the energy enclosure.

This approach allows us to measure estimate L and H based on the measurements of dry- and wet-bulb temperatures at two heights for L and H, avoiding direct measurements of vapor density in the air (Rosenberg *et al.* 1983). The Bowen ratio method has been widely used to estimate evapotranspiration (ET) prior to the eddy-covariance method (Chapter 4).

1.8 Summary

Popular environmental biophysical models applied in ecosystem studies are largely developed based on several key physical processes (*e.g.*, radiation and kinetic energy flows) near the land surface. Understanding these processes is essential before discussing the other models presented in the following chapters. The comprehensive description of these processes, however, is a large, multi-disciplinary field of study. The several major processes that are presented in this chapter were selected because of their foundational roles in constructing, understanding and applying the models that will be introduced in later chapters, but this chapter is far from a comprehensive review of our current knowledge on environmental biophysics. Several classical texts are thus provided as must-read references for advanced users.

Similarly, I have focused on the diel changes of the aforementioned variables and parameters in this chapter largely because the dominant models that will be introduced in later chapters are applied at the hourly scale. I recognize that not all ecosystem models or Earth system models are designed to run at an hourly scale. The purpose here is to help users to understand the theoretical (or empirical) foundations of different physical processes and relationships among the variables, though additional measures are needed at broader temporal scales (month, season, and years).

To better understand the underlying biophysical principles and processes for the diel changes of temperature, vapor pressure, VDP, heat storage and fluxes in soils, solar radiation, wind profile, and energy fluxes near the land surface, hourly data collected on September 22, 2016, from a switchgrass field at the W. K. Kellogg Biological Station of Michigan State University are used to demonstrate the models' behaviors and capabilities. A complete dataset of continuous records of 84 variables from this site are included in the appendix of this chapter through the book's Web connection. Four Excel spreadsheet models (coupled with graphic displays) and one Python calculator are provided for such a purpose. While these models can be applied elsewhere with site-specific input information, users should be cautious about model assumptions and requirements. The International System of Units (SI) is applied, and a consistent use of symbols is maintained whenever possible.

Online Supplementary Materials

S1-1: Micrometeorological data collected at an open-path eddy-covariance tower located in a switchgrass cropland of the W. K. Kellogg Biological Station (KBS), Michigan, USA, in 2016 (Switchgrass_metdata2016.xlsx). This dataset is provided for modeling exercises (e.g., S1-2, S1-3, S1-4), as well as for model parameterizations in Chapters 2 – 5.

S1-2: A simple simulator of diel change of air temperature based on daily maximum and minimum temperature, as well as their timings (Ta_Diel.xlsx).

S1-3: Calculations of wet-bulb temperature (T_w), dew point temperature (T_d), vapor pressure (e_a), saturation vapor pressure (e_s) and vapor pressure deficit (VPD) from *in-situ* measurements of air temperature (T_a) and relative humidity (h) (Eqs. 1.5 – 1.13) (Ta_h_VPD.xlsx).
S1-4: An empirical model for simulating vertical profile of wind speed over a hypothetical vegetation based on Equation (1.24) (Wind.xlsx).
S1-5: Python codes for simulating the position of the Sun over a given location. Input variables include latitude, longitude, elevation (m), year, month and day of year; outputs are solar zenith angle, declination and day length at 0.1-hour interval (Solar.py).

Scan the QR code or go to
https://msupress.org/supplement/BiophysicalModels
to access the supplementary materials.
User name: biophysical
Password: z4Y@sG3T

Acknowledgements

The author appreciates Michael Abraha for sharing his field data from an eddy-covariance tower, Ariclenes Silva and Huimin Zou for recoding my original C^{++} codes for simulating the Sun movement into Python codes, and Housen Chu and Ankur Desai for detailed suggestions on early drafts of the chapter. Kristine Blakeslee edited the language and format of the draft.

References

Abraha, M., Chen, J., Chu, H., Zenone, T., John, R., Su, Y. J., Hamilton, S. K., and Robertson, G. P. (2015). Evapotranspiration of annual and perennial biofuel crops in a variable climate. *Global Change Biology-Bioenergy*, 7(6), 1344–1356.

Abraha, M., Hamilton, S. K., Chen, J., and Robertson, G. P. (2018). Ecosystem carbon exchange on conversion of Conservation Reserve Program grasslands to annual and perennial cropping systems. *Agricultural and Forest Meteorology*, 253, 151–160.

Baldocchi, D., Falge, E., Gu, L., Olson, R., Hollinger, D., Running, S., Anthoni, P., Bernhofer, C., Davis, K., Evans, R., Fuentes, J., Goldstein, A., Katul, G., Law, B., Lee, X., Malhi, Y., Meyers, T., Munger, W. Oechel, W., Paw U, K. T., Pilegaard, K., Schmid, H. P., Valentini, R., Verma, S., Vesala, T., Wilson, K., and Wofsy, S. (2001).

FLUXNET: A new tool to study the temporal and spatial variability of ecosystem-scale carbon dioxide, water vapor, and energy flux densities. *Bulletin of the American Meteorological Society*, 82(11), 2415–2434.

Bonan, G. (2019). *Climate Change and Terrestrial Ecosystem Modeling.* Cambridge University Press. 459pp.

Campbell, G. S., and Norman, J. (2012). *An Introduction to Environmental Biophysics.* Springer Science & Business Media. 286pp.

Chen, J., Franklin, J. F., and Spies, T. A. (1993a). An empirical model for predicting diurnal air-temperature gradients from edge into old-growth Douglas-fir forest. *Ecological Modelling*, 67(2-4), 179–198.

Chen, J., Franklin, J. F., and Spies, T. A. (1993b). Contrasting microclimates among clearcut, edge, and interior of old-growth Douglas-fir forest. *Agricultural and Forest Meteorology*, 63(3-4), 219–237.

Chen, J., Saunders, S. C., Crow, T. R., Naiman, R. J., Brosofske, K. D., Mroz, G. D., Brookshire, B. L., and Franklin, J. F. (1999). Microclimate in forest ecosystem and landscapes. *BioScience*, 49(4), 288–297.

Chu, H., Baldocchi, D. D., Poindexter, C., Abraha, M., Desai, A. R., Bohrer, G., Arain, M. A., Griffis, T., Blanken, P. D., O'Halloran, T. L., Zhang, Q., Burns, S., Frank, J. M., Christian, D., Brown, S., Black, T. A., Gough, C. M., Law, B. E., Lee, X., Chen, J., Reed, D. E., Massman, W. J., Clark, K., Hatfield, J., Prueger, J., Bracho, R., Baker, J. M., Martin, T. A., and Thomas, R. Q. (2018). Temporal dynamics of aerodynamic canopy height derived from eddy covariance momentum flux data across North American flux networks. *Geophysical Research Letters*, 45(17), 9275–9287.

Cleverly, J., Thibault, J. R., Teet, S. B., Tashjian, P., Hipps, L. E., Dahm, C. N., and Eamus, D. (2015). Flooding regime impacts on radiation, evapotranspiration, and latent energy fluxes over groundwater-dependent riparian cottonwood and saltcedar forests. *Advances in Meteorology*, 2015:935060. DOI: 10.1155/2015/935060.

Dewar, R. C. (2002). The Ball-Berry-Leuning and Tardieu-Davies stomatal models: Synthesis and extension within a spatially aggregated picture of guard cell function. *Plant, Cell & Environment*, 25(11), 1383–1398.

Ebi, K. L., Burton, I., and McGregor, G. (2009). *Biometeorology for Adaptation to Climate Variability and Change.* Springer, Dordrecht. 281pp.

Fritschen, L. J., and Gay, L. W. (1979). *Environmental Instrumentation.* Springer Science & Business Media. 216pp.

Griffiths, J. F. (1994). *Handbook of Agricultural Meteorology.* Oxford University Press. 320pp.

Goulden, M. L., Munger, J. W., Fan, S. M., Daube, B. C., and Wofsy, S. C. (1996). Measurements of carbon sequestration by long-term eddy covariance: Methods and a critical evaluation of accuracy. *Global Change Biology*, 2(3), 169–182.

Gu, L., Falge, E. M., Boden, T., Baldocchi, D. D., Black, T. A., Saleska, S. R., Suni, T., Verma, S. B., Vesala, T., Wofsy, S. C., and Xu, L. (2005). Objective threshold determination for nighttime eddy flux filtering. *Agricultural and Forest Meteorology*, 128(3-4), 179–197.

Kaimal, J. C., and Finnigan, J. J. (1994). *Atmospheric Boundary Layer Flows: Their Structure and Measurement.* Oxford University Press. 289pp.

Kimball, B. A., and Bellamy, L. A. (1986). Generation of diurnal solar radiation, temperature, and humidity patterns. *Energy in Agriculture*, 5(3), 185–197.

Lee, X., Massman, W., and Law, B. (2004). *Handbook of Micrometeorology: A Guide for Surface Flux Measurement and Analysis (Vol. 29). Springer Science & Business Media. 250pp.*

Lindroth, A., Mölder, M., and Lagergren, F. (2010). Heat storage in forest biomass improves energy balance closure. *Biogeosciences*, 7(1), 301–313.

Lowry, W. P. (2013). Weather and Life: An Introduction to Biometeorology. Elsevier. 305pp.Monteith, J., and Unsworth, M. (2013). *Principles of Environmental Physics: Plants, Animals, and the Atmosphere.* Academic Press. 423pp.

Oliphant, A. J., Grimmond, C. S. B., Zutter, H. N., Schmid, H. P., Su, H. B., Scott, S. L., Offerle, B. J. C. R., Randolph, J. C., and Ehman, J. (2004). Heat storage and energy balance fluxes for a temperate deciduous forest. *Agricultural and Forest Meteorology*, 126(3-4), 185–201.

Papale, D., Reichstein, M., Aubinet, M., Canfora, E., Bernhofer, C., Kutsch, W., Longdoz, B., Rambal, S., Valentini, R., Vesala, T., and Yakir, D. (2006). Towards a standardized processing of net ecosystem exchange measured with eddy covariance technique: Algorithms and uncertainty estimation. *Biogeosciences,* 3, 571–583.

Parton, W. J., and Logan, J. A. (1981). A model for diurnal variation in soil and air temperature. *Agricultural Meteorology*, 23, 205–216.

Pennypacker, S., and Baldocchi, D. (2016). Seeing the fields and forests: Application of surface-layer theory and flux-tower data to calculating vegetation canopy height. *Boundary-layer Meteorology*, 158(2), 165–182.

Reichstein, M., Falge, E., Baldocchi, D., Papale, D., Aubinet, M., Berbigier, P., Bernhofer, C., Buchmann, N., Gilmanov, T., Granier, A., Grünwald, T., Havrávanková, K., Ilvesniemi, H., Janous, D., Knohl, A., Laurila, T., Lohila, A., Loustau, D., Matteucci, G., Meyers, T., Miglietta, F., Ourcival, J-M., Pumpanen, J., Rambal, S., Rotenberg, E., Sanz, M., Tenhunen, J., Seufert, G., Vaccari, F., Vesala, T., Yakir, D., Valentini, R., and Grünwald, T. (2005). On the separation of net ecosystem exchange into assimilation and ecosystem respiration: Review and improved algorithm. *Global Change Biology*, 11(9), 1424–1439.

Rosenberg, N. J., Blad, B. L., and Verma, S. B. (1983). *Microclimate: The Biological Environment.* John Wiley & Sons. 495pp.

Shao, C., Chen, J., Li, L., Dong, G., Han, J., Abraha, M., and John, R. (2017). Grazing effects on surface energy fluxes in a desert steppe on the Mongolian Plateau. *Ecological Applications*, 27(2), 485–502.

Shao, C., Chen, J., Li, L., Xu, W., Chen, S., Gwen, T., Xu, J., and Zhang, W. (2008). Spatial variability in soil heat flux at three Inner Mongolia steppe ecosystems. *Agricultural and Forest Meteorology*, 148(10), 1433–1443.

Zenone, T., Chen, J., Deal, M. W., Wilske, B., Jasrotia, P., Xu, J., Bhardwaj, A. K., Hamilton, S. K., and Robertson, G. P. (2011). CO_2 fluxes of transitional bioenergy crops: Effect of land conversion during the first year of cultivation. *Global Change Biology-Bioenergy*, 3(5), 401–412.

Chapter 2
Modeling Ecosystem Production

Jiquan Chen

2.1 Introduction

Ecosystem production is the amount of organic compound generated in a land area. Gross primary production (GPP) is the amount of organic substances produced by plants through photosynthesis (P_n, or A_n). Plants need carbon dioxide (CO_2), water (H_2O), energy and nutrients to conduct photosynthesis, transport water from soil to leaves (*i.e.*, transpiration), and convert and transport synthesized carbon to different parts of organs. The amount of carbon needed to maintain photosynthesis and carbon reallocations is called autotrophic respiration (R_a). The difference between GPP and R_a is termed net primary production (NPP). Dead materials from plants are broken down into simple organic substances by microorganisms, and eventually released as gases (mostly as CO_2) back into the atmosphere. This process is called heterotrophic respiration (R_h). The sum of R_a and R_h equals ecosystem respiration (R_{eco}). The difference between NPP and R_h is net ecosystem production (NEP). The relationship among these flux terms is expressed in Figure 2.1.

There are other minor components of carbon cycles in an ecosystem. These have been well described in many textbooks in ecology, forestry and

---

Jiquan Chen
Landscape Ecology & Ecosystem Science (LEES) Lab, Department of Geography, Environment, and Spatial Sciences & Center for Global Change and Earth Observations, Michigan State University, East Lansing, MI 48823
Email: jqchen@msu.edu

Jiquan Chen, *Biophysical Models and Applications in Ecosystem Analysis*,
https://doi.org/10.3868/978-7-04-055256-0-2

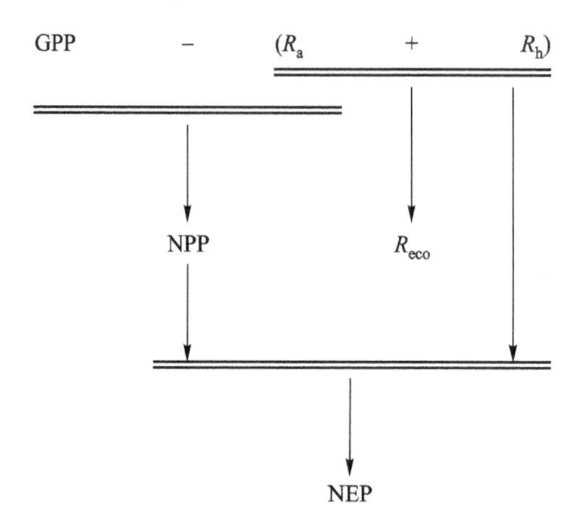

Fig. 2.1 Schematic illustration of the major carbon fluxes for ecosystem production.

agriculture (*e.g.*, Waring and Running 2010, Chapin *et al.* 2011, Chen *et al.* 2014). Modeling respiration will be covered in Chapter 3, while this chapter focuses on GPP, which is determined by photosynthesis. Production is conventionally expressed in units of mass per unit area per unit time. In ecosystem studies, mass of carbon per unit area per year (g C m^{-2} yr^{-1}) is most often used as the unit of production, although other expressions are common in the literature, including mg C m^{-2} s^{-1}, µmol CO_2 m^{-2} s^{-1}, g C m^{-2} yr^{-1}, Mg C ha^{-1} yr^{-1}. The molar mass of CO_2 is 44 g mol^{-1}, with 27.3% as carbon (12 g mol^{-1}). The molar mass of water (H_2O) is ~18 g mol^{-1}, with 11.2% as hydrogen (H) that has an atomic mass of 1. Organic material (*i.e.*, biomass) is expressed in dry weight per unit land area, Mg ha^{-1} (*i.e.*, t ha^{-1}) or annual production, Mg ha^{-1} yr^{-1}. For terrestrial ecosystems, approximately 50% of biomass is considered to be carbon, albeit this can vary from 45.0% to 47.9% among terrestrial plants (Ma *et al.* 2018).

Predicting ecosystem productivity has always been a core effort in all disciplines of natural science. In forestry, a central focus has been on timber production, which is commonly called growth and yield modeling (Clutter 1983). Stand age, density, site index, diameter, crown ratio, and tree height are used to construct growth-yield tables with forest type, stand age, site index, and management activities (*e.g.*, thinning, harvesting, rotation, *etc.*). For agricultural crops, the focus has been on grain production, which is modeled using crop type, weather and climatic conditions (*e.g.*, growing degree-days), soil, and management (*e.g.*, fertilization, irrigation, weed controls, *etc.*) (Evans 1996). In grassland ecology, type, canopy cover, height, climate, and human disturbances have been used as the primary predictors for grass production (Parton 1996). The above mentioned models rely on *in-situ* measurements and are empirically developed with ground measurements of both production and independent variables. Various forms have been proposed for constructing these empirical models (*e.g.*, Landsberg 1977).

With the installation of the International Geosphere-Biosphere Programme (IGBP) in the late 1960s, process-based models emerged (Schlesinger 1977). The JABOWA (Botkin *et al.* 1972) and FORET class models (*a.k.a.* gap models) are among the earliest examples that considered interactions among trees for modeling forest production and dynamics, with localized climates as key drivers (Shugart 1984). Since the 1980s, detailed ecosystem processes have been included in dozens of ecosystem models to address ecosystem production under changing climate and human disturbances. These models were designed for broad applications (*e.g.*, Schaefer *et al.* 2012) and emphasize physiology (*e.g.*, Landsberg and Sands 2011, Thornley and Johnson 1990.), soil (*e.g.*, CENTURY model) (Parton 1996), energy and mass balance (SiB models) (Sellers *et al.* 1986), species interactions (Shugart 1984), resource use (PnET) (Aber and Federer 1992), or management (*e.g.*, 3PG model) (Landsberg and Waring 1997). Nevertheless, most models include critical biophysical processes (*e.g.*, photosynthesis, respiration, nutrient cycling, and genetics) with often one or more temporal, spatial or organismal resolutions left unspecified (Muller and Martre 2019).

Photosynthesis is the first step for assimilating atmospheric CO_2 into organic substances in an ecosystem and is the focus of this chapter, so readers can understand the key algorithms employed in most ecosystem models. Photosynthesis is a physiological process in which plants, algae and certain bacteria convert solar energy and CO_2 to chemical energy and carbohydrate — such as glucose, sugar, and cellulose. The name "photosynthesis" is a combination of the Greek words "light" and "putting together". The process was discovered by Dutch physician Jan Ingenhousz in the late 1700s. Photosynthesis involves many complex chemical and biological processes and is performed differently by different functional groups of species. There are a number of pigments involved, but all the chemical conversions take place with chlorophyll a. Two types of chlorophyll pigments absorb light in the blue and red part of the visible spectrum. Its chemical expression has several forms, including:

$$6CO_2 + 12H_2O + \text{Solar Energy} \rightarrow C_6H_{12}O_6 + 6O_2 + 6H_2O$$

where six molecules of CO_2 combine with 12 molecules of H_2O using light to split water. The detailed chemical processes is very similar in most photosynthetic organisms but there are some variations in how CO_2 is taken up, for example, C_3 plants (diffusion only) and C_4 plants (active CO_2 accumulation) (Ehleringer and Björkman 1977). The solar energy used in photosynthesis is derived from photosynthetically active radiation (PAR) in the visible wavelengths from 400 – 700 nm (*see* Chapter 5) because visible light is not equally used by pigments. The product of photosynthesis is the formation of sugars along with one molecule of breakable oxygen per carbon fixed and water. Based on this principle, photosynthesis or gross production of an ecosystem is modeled with leaf mass (*e.g.*, leaf area and leaf mass), radiation energy (*e.g.*, PAR), availability of soil water and nutrients, as well as climatic conditions (Section 2.2).

Since plant or ecosystem production requires CO_2, H_2O, PAR, chlorophylls (leaves), and essential macro- and micro-nutrients, the amount and availability of these resources dictate the kind of algorithms selected in modeling production. This is also known in agronomy as the "G x E x M" concept of crop production (*e.g.*, grain yield), and it is a function of genetics (G), environment (E), and management (M) (*e.g.*, Montesino-San Martín *et al.* 2014; Muller and Martre 2019). Recent synthesis of global data also confirms that plant functional type, climate and disturbances (both natural and human) are key variables for predicting the magnitude of ecosystem production and its change over time (Peaucelle *et al.* 2019).

Before any model can be widely used it must first be parameterized and tested. The data needed for model parameterizations and validations come in different forms. Each has advantages and limitations, and are appropriate for specific temporal and spatial scales. Ground-based measurements usually do not tally the contributions of individual species but measurement of net exchange of CO_2 provides direct values of production, whereas remote sensing-based approaches provide indirect estimates across continuous space but are limited by their sampling frequency and atmospheric conditions (Chen *et al.* 2004).

In the field, data are collected with chambers (Pearcy *et al.* 2012) such as the LI-COR Portable Photosynthesis System (Fig. 2.2a) that dominates photosynthesis studies (*e.g.*, Ehleringer and Cook 1980, Hunt 2003). These systems have been modified to measure photosynthesis of branches and small plants. Rapid development in the eddy-covariance approach (Lee *et al.* 2004) provides measurements of net exchange of CO_2, H_2O, energy, trace gases, and microclimate over homogeneous canopies (Fig. 2.2b). Both chamber-based and micrometeorology-based methods provide instantaneous measurement of carbon assimilation and losses (Fig. 2.2a, b). FLUXNET — a collaborative association of regional networks — houses thousands of year-site data on fluxes of trace gases between ecosystems and the atmosphere[1]. These measurements represent ecosystem level NEP and can be partitioned into GPP and R_{eco} (Reichstein *et al.* 2005) (*see* notes in Section 2.2.1). Stable carbon and oxygen isotopic analyses are increasingly used to tease apart different components of production such as autotrophic and heterotrophic respiration (*e.g.*, Farquhar *et al.* 1989, Yakir and Sternberg 2000, Pataki *et al.* 2003, Helliker *et al.* 2018).

Fig. 2.2 Examples of various ways to assess components of ecosystem production: (a) with a portable photosynthesis chamber; (b) an open path eddy-covariance flux tower; (c) via biometric sampling of diameter at breast height (DBH) and stem mapping; and (d) using lidar images of canopy height and tree distribution (modified from Giannico *et al.* 2016).

Biometric measurements of ecosystem production using conventional plot sampling methods are another major data source. Community composition (*e.g.*, species), structure (diameter, canopy cover, height, leaf area index (LAI), coordinates of individual plants) and biomass are measured sequentially at 1 – 5 year intervals (Fig. 2.2c). The change in above- and below-ground biomass between two consecutive years is NEP, assuming both below- and above-ground assessments are made. The monitoring plots at the Long-

1 https://fluxnet.fluxdata.org/

Term Ecological Research (LTER)[1] sites (Franklin *et al.* 1990), for example, have data since 1980s that can be used for calculating NEP or NPP (*e.g.*, Knapp *et al.* 2001). Similar data are also increasingly developed and available through other coordinated networks (*e.g.*, CERN, NEON, LTAR, *etc.*). For temperate forests, diameter growth of trees is well preserved in tree rings, allowing direct assessment of historical changes of stem growth (*i.e.*, NEP of the stems) and indirectly estimate NEP. The International Tree-Ring Data Bank (ITRDB)[2] is the world's largest public archive of tree ring data, including raw ring width, wood density, and isotope measurements, site index, *etc.* The Laboratory of Tree-Ring Research (LTRR)[3], established in 1937 at the University of Arizona, is another source of data in Tucson, Arizona, USA. These data are valuable for modeling relative changes in above-ground production in response to major regulatory variables (*e.g.*, LAI, PAR, soil moisture) (Bhuyan *et al.* 2017) or disturbances (*e.g.*, insect defoliation, heat waves, wild fires, *etc.*). Another useful data source is the plant phenology that has been increasingly recognized as a critical variable in modeling ecosystem production. The USA National Phenology Network (USA-NPN)[4] was established in 2007 to collect, store, and share phenological data and other information. Finally, climatic variables are also needed in modeling ecosystem production. These data are now widely organized for public use by many countries and organizations (*e.g.*, Food and Agricultural Organization)[5].

2.2 Core Biophysical Models for Ecosystem Production

Models to be covered in this chapter can be divided into three broad types:

1. Modeling photosynthesis rates (P_n, or A_n) from PAR (*i.e.*, light response curve), CO_2 concentration (*i.e.*, A-c_i curve), and other biophysical variables (*e.g.*, VPD, temperature, soil moisture, leaf quantity and quality).
2. Modeling gas exchange of both CO_2 and water vapor through stomatal conductance (g_s).

---

1 https://lternet.edu/
2 https://www.ncdc.noaa.gov/data-access/paleoclimatology-data/datasets/tree-ring
3 https://ltrr.arizona.edu/
4 https://www.usanpn.org/
5 http://www.fao.org/3/X0490E/x0490e07.htm

3. Modeling gross primary production (GPP) through constraints imposed by limiting resources supplies (*e.g.*, light, nutrient, water, and carbon-use efficiency).

In the next section, a number of different mathematical models will be presented that are designed to predict gross or net photosynthesis. Near the end of the chapter, a comparison of model performance will be made against carefully acquired field data. For more detailed information, readers are encouraged to access the original publications where models were derived (*e.g.*, Monteith 1972, Farquhar *et al.* 1980, Shugart 1984, Ball *et al.* 1987, Thornley and Johnson 1990, Aber and Federer 1992, Xiao *et al.* 2004, Sharkey *et al.* 2007, Farquhar *et al.* 2001, von Caemmerer 2013, Bonan *et al.* 2014, Buckley 2017).

2.2.1 Michaelis-Menten Model

Photosynthesis rate (P_n) increases initially with PAR at a variable rate (α). The increasing trend will level off as light is saturated (*i.e.*, $\alpha = 0$), eventually resulting in the maximum photosynthesis (P_m). These changes in photosynthesis with PAR have been widely modeled by using Michaelis-Menten (MM) kinetics (Michaelis and Menten 1913). MM kinetics describes a specific case of enzymes. It can be applied to many things because it describes a rectangular hyperbola (Chapter 3). The relationship between P_n and PAR, also commonly known as light response curve, is expressed as a rectangular hyperbolic model:

$$P_n = \frac{\alpha \cdot \text{PAR} \cdot P_m}{\alpha \cdot \text{PAR} + P_m} \tag{2.1}$$

where α is the photochemical efficiency of photosynthesis at low light and P_m (μmol m^{-2} s^{-1}) is the maximum photosynthetic capacity of a leaf or an ecosystem. A unique property of this model is that the Michaelis constant (K_m) of the enzyme is an inverse measure of affinity. K_m is the value when P_n reaches half of the P_m. Both P_m and K_m values help a user to evaluate and compare the light response curves among leaves or ecosystems. The Michaelis-Menten model (Eq. 2.1), however, assumes that the photosynthetic rate is zero when PAR = 0. In reality, measured net exchange of CO_2 through stomata (or an ecosystem) is the difference between gross photosynthesis and respiration (R_d) — the simultaneous CO_2 loss during photosynthesis, including growth and maintenance costs (Thornley 2011).

This causes a P_n of zero until PAR reaches a certain level (*i.e.*, light compensation point, I_o). Inclusion of R_d (or R_e) in Equation (2.1) is necessary:

$$P_n = \frac{\alpha \cdot \text{PAR} \cdot P_m}{\alpha \cdot \text{PAR} + P_m} - R_d \tag{2.2}$$

Equation (2.2) assumes the light response curve is vertically lowered by R_d (or R_e). An alternative is to replace P_m with $(P_m - R_d)$, but the interpretations of the model parameters would need to be adjusted accordingly. It is worth noting that daytime respiration is proportional to that recorded at comparable temperatures at night (Reichstein *et al.* 2005) — a common approach applied with the FLUXNET community to calculate GPP. This assumption is questionable because photosynthesis and the related exudate production can affect soil microbial activity (Högberg *et al.* 2001).

While the Michaelis-Menten equation has a unique feature determined by the Michaelis constant, the shape of the light response curve is not flexible. Landsberg and Sands (2011) introduced an additional shape factor (β) into a non-rectangular hyperbolic model (*see also* Buckley and Diaz-Espejo 2015):

$$P_n = P_m \cdot \frac{2 \cdot \alpha \cdot \text{PAR}/p_m}{1 + \alpha \cdot \frac{\text{PAR}}{P_m} + \sqrt{\left(1 + \alpha \cdot \frac{\text{PAR}}{P_m}\right)^2 - 4 \cdot \alpha \cdot \beta \cdot \text{PAR}/P_m}} \tag{2.3}$$

This model is virtually the same as Equation (2.1) when $\beta = 0$. The value of β should be less than 1 for simulations.

An alternative expression of the non-rectangular hyperbolic model is applied by Peat (1970) as:

$$P_n = \frac{1}{2 \cdot \beta} \left[\alpha \cdot \text{PAR} + P_m - \sqrt{(\alpha \cdot \text{PAR} + P_m)^2 - 4 \cdot \alpha \cdot \text{PAR} \cdot P_m \cdot \beta}\right] \tag{2.4}$$

Gilmanov *et al.* (2003) point out that the rectangular hyperbola tends to overestimate the initial slope (α) and P_m, and the nonrectangular model provides a much better fit (Eq. 2.1), albeit more measurements are needed to have large sample size because extra parameters are involved in non-linear regression analysis.

Six sets of parameters are used to illustrate model behaviors (S2-1: LightResponse.xlsx). With a P_m of 10 (μmol m^{-2} s^{-1}), P_n (μmol m^{-2} s^{-1}) increases faster with a higher α value and eventually moves toward its capacity (P_m) (scenarios 1–3). The Michaelis constant is 83, 200, 500 (μmol m^{-2} s^{-1}), respectively, in scenarios 1 – 3. The curves are lowered with reduced P_m values (scenarios 4 and 5) or with R_d included (scenario 6) (Fig. 2.3a). For the hyperbolic model (Eq. 2.3), a β value of < 1 will result in a faster increase in P_n with PAR (μmol m^{-2} s^{-1}), whereas negative values will lower the increasing rate of P_n with PAR (Fig. 2.3b). A user is encouraged to test different sets of parameters for the models before performing the corresponding non-linear regression analysis.

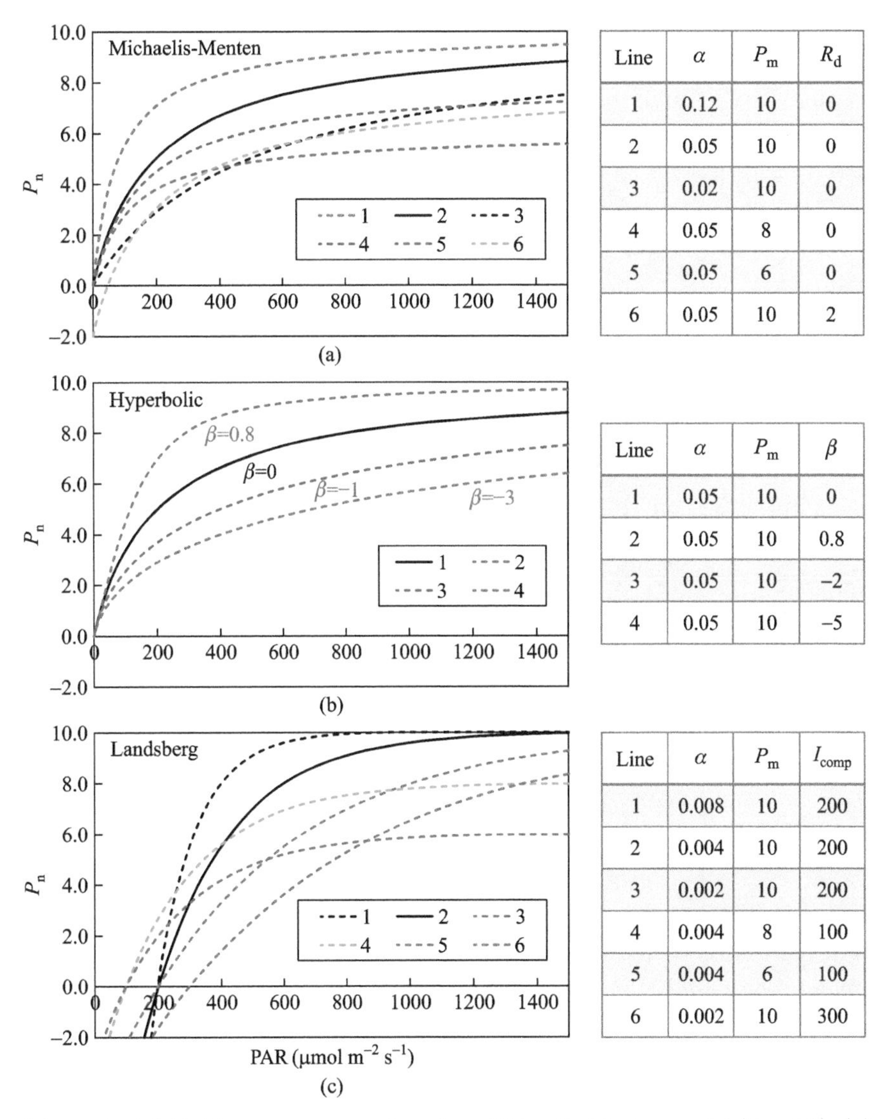

Line	α	P_m	R_d
1	0.12	10	0
2	0.05	10	0
3	0.02	10	0
4	0.05	8	0
5	0.05	6	0
6	0.05	10	2

Line	α	P_m	β
1	0.05	10	0
2	0.05	10	0.8
3	0.05	10	−2
4	0.05	10	-5

Line	α	P_m	I_{comp}
1	0.008	10	200
2	0.004	10	200
3	0.002	10	200
4	0.004	8	100
5	0.004	6	100
6	0.002	10	300

Fig. 2.3 (a) Light response curves based on Michaelis-Menten model (Eq. 2.2); (b) non-rectangular hyperbolic model; and (c) Landsberg model (Eq. 2.5) with different sets of model parameters. Note that modeled curve with $\beta = 0$ in (b) is the same as scenario 2 in (a). Other curves can be generated by altering parameters in S2-1 (LightResponse.xlsx).

2.2.2 Landsberg Model

One of the models proposed by Landsberg (1977) for biological growth processes is the saturating exponential model. When applied for light-response curve, it is expressed as:

$$P_n = P_m \cdot \left[1 - e^{\alpha \cdot (PAR - I_{comp})}\right] \tag{2.5}$$

where P_m (μmol m^{-2} s^{-1}) is the maximum photosynthesis, α is the slope of the change in P_n (μmol m^{-2} s^{-1}) with PAR (*i.e.*, a shape factor), and I_{comp} (μmol m^{-2} s^{-1}) is the light compensation point at which P_n is zero (also labeled as I_0). This model assumes that α value is a constant. Inclusion of I_{comp} in the exponential model provides us a unique feature: P_n switches from negative to positive (*i.e.*, gross photosynthesis minuses photorespiration) at a certain PAR level. The equations were designed for leaf-level photosynthesis but have been applied to describing ecosystem level CO_2 fluxes as a function of irradiance (Hollinger *et al.* 1994, Chen *et al.* 2002). The Landsberg model emphasizes the statistical properties of the light response curve (*i.e.*, compensation and saturation point), whereas the Michaelis-Menten model (Eqs. 2.1 – 2.4) was formed with enzyme kinetics. Six simulations are included in the demonstrative spreadsheet (S2-1). Again, a user can explore the light responses by providing different values of α, P_m, and I_{comp} (Fig. 2.3c).

The rectangular and non-rectangular approach were attempted in 1950s (*see* Thornley and Johnson 1990). Light response models are often constructed with *in-situ* measurements at leaf, individual or ecosystem level. When the same model is applied with empirically estimated coefficients, one can infer the underlying processes. At leaf level, for example, one can assess differences among leaves of different species, at different positions in the canopy, during different seasons, or under different climatic and soil conditions. At ecosystem level, the models have been applied to fill data gaps to create continuous time series, examine ecosystem responses to changes in local climate and disturbances, scale up from individual ecosystems to landscape-region-global levels, and investigate other biophysical regulations such as fertility through quantum efficiency, VPD, temperature constraints, phenology, soil moisture and nutrients, disturbances, *etc.*

2.2.3 Farquhar's Model

Photosynthesis rate is not only regulated by radiation (*i.e.*, PAR) and water supplies from the soil (*see* Section 2.1) but also by the intercellular CO_2 concentration (*i.e.*, CO_2 supply) that depends on CO_2 diffusion through stomata (*i.e.*, stomatal conductance, g_s). Farquhar, von Caemmerer and Berry (1980) proposed a biochemical growth model for leaf level photosynthesis rate, known as Farquhar's model. There are several ways to express the model in the literature (note that different sets of symbols are used in this section to be consistent with those used in plant physiology). Net leaf CO_2 assimilation (A_n, μmol m^{-2} s^{-1}) is the least of the three rates:

$$A_n = \text{minimum}\,(A_c, A_j, A_p) \tag{2.6}$$

where A_c, A_j, and A_p are the photosynthesis rate for Rubisco-limited, RuBP-limited, and product-limited assimilations, respectively. Photosynthesis rate as a function of intercellular CO_2 concentration is described by FvCB equation:

$$A_c = \frac{V_{max} \cdot (c_i - \Gamma^*)}{c_i + K_c \cdot \left(1 + \frac{o_i}{K_o}\right)} \tag{2.7}$$

where V_{max} is the maximum activity of Rubisco, c_i is the intercellular CO_2 concentration (μmol mol^{-1}), Γ^* is the CO_2 compensation point in the absence of day respiration (R_d), K_c is the Michaelis-Menten constant of Rubisco for CO_2, O_i is the oxygen (O_2) concentration in the atmosphere (209 mol mol^{-1}), and K_o is the Michaelis-Menten constant of Rubisco for O_2. Γ^* is calculated as:

$$\Gamma^* = \frac{0.5 \cdot O_i}{2600 \cdot 0.57^{Q_{10}}} \tag{2.8}$$

where Q_{10} is the leaf respiration and can be modeled with leaf temperature (Evans 1987). Details about Q_{10} modeling are provided in Chapter 3. The dependence of quantum yield on wavelength and growth irradiance. K_c for CO_2 is calculated as:

$$K_c = 30 \cdot 2.1^{Q_{10}} \tag{2.9}$$

and K_o for O_2 is calculated as:

$$K_o = 30000 \cdot 1.2^{Q_{10}} \tag{2.10}$$

V_{max} as the maximum capacity of Rubisco varies with leaf temperature, foliar carbon and nitrogen ratio, soil moisture, and other biophysical conditions (Wolf *et al.* 2006).

RuBP-limited photosynthesis rate (A_j), also commonly known as light-limited photosynthesis rate, is calculated as:

$$A_j = \frac{j \cdot (c_i - \Gamma^*)}{4 \cdot c_i + 8 \cdot \Gamma^*} \quad (2.11)$$

where j is the electron transport rate (μmol m^{-2} s^{-1}) and varies with absorbed photosynthetically active radiation (aPAR). Finally, the product-limited photosynthesis rate is calculated as:

$$A_p = 3 \cdot T_p \quad (2.12)$$

where T_p (μmol m^{-2}) is the triose phosphate utilization rate. This rarely limits the rate of photosynthesis under physiological conditions (Kumarathunge *et al.* 2019) but inclusion in the models improves parameterizations (Sharkey 2019).

The four major parameters that are needed to fit Farquhar's model through constructions of review for the A–c_i curve are V_{max} (μmol m^{-2} s^{-1}), J_{max} (μmol m^{-2} s^{-1}), T_p (μmol m^{-2} s^{-1}), and R_d (μmol m^{-2} s^{-1}). Sharkey *et al.* (2007) (*see also* Sharkey 2015) provided a comprehensive review of the models and their applications. There are a number of updated methods for estimating the parameters in Equations (2.7)–(2.11), including those described by Medlyn *et al.* (2002), Kattge and Knorr (2007), Bonan *et al.* (2014), and Moualeu-Ngangue *et al.* (2017). There also are many online tools that provide parameter estimations and model demonstrations, and some of these are equipped with original codes, principles of Farquhar's model, and interactive illustrations. The Denning Lab at Colorado State University is one resource for model demonstration[1]. Bellasio *et al.* (2019) presented more sophisticated fitting tools in Excel. The LeafWeb provides automated numerical analyses of leaf gas exchange measurements[2].

2.2.4 Photosynthesis based on Stomatal Conductance (g_s)

Carbon dioxide for photosynthesis diffuses into intercellular chambers through leaf stomata. The diffusion rate is called stomatal conductance (g_s, μmol

1 https://biocycle.atmos.colostate.edu/shiny/photosynthesis/
2 https://leafweb.org/

m^{-2} s^{-1}), which is proportional to the photosynthesis rate (A_n, μmol m^{-2} s^{-1}). This linear relationship is modulated by leaf surface CO_2 and H_2O concentration and varies among leaves and species. Understanding the changes in g_s is a powerful tool not only for predicting photosynthesis by reversing the $[g_s \sim A_n]$ relationship but also for modeling transportation of other materials through stomata (*e.g.*, H_2O-transpiration). Ball *et al.* (1987) proposed a simple empirical model for the diurnal changes of g_s for simulating the exchanges of CO_2 and H_2O at leaf surface, known as the Ball-Berry model:

$$g_s = K \cdot A_n \cdot \frac{h_s}{c_s} \tag{2.13}$$

where h_s (ranging 0 – 1) is the fractional relative humidity at the leaf surface, c_s (μmol mol^{-1}) is the CO_2 concentration of leaf surface, and K is the slope constant of the model that represents the composite sensitivity of g_s to CO_2 concentration. By reversing Equation (2.13), photosynthesis is modeled as:

$$A_n = \frac{g_s \cdot c_s}{K \cdot h_s} \tag{2.14}$$

Because stomata do not completely close, there is a minimum conductance value (g_0, mol m^{-2} s^{-1}). The Ball-Berry model is also expressed as:

$$g_s = g_0 + g_1 \cdot A_n \cdot \frac{h_s}{c_s} \tag{2.15}$$

where g_0 and g_1 are empirically estimated intercept and slope, respectively. By reversing Equation (2.15), we have:

$$A_n = \frac{(g_s - g_0) \cdot c_s}{g_1 \cdot h_s} \tag{2.16}$$

The challenge here is to measure or estimate stomatal conductance (g_s) under different conditions (*e.g.*, PAR, temperature and vapor pressure, soil moisture and fertility), as well as over time and among species.

Leuning (1990) argued that the use of $(c_s - \Gamma)$ is more appropriate in the numerator, and he modified the original Ball-Berry model:

$$g_s = g_0 + \frac{a_1 \cdot A_n}{c_s - \Gamma} \tag{2.17}$$

Leuning reasoned this new form was applicable because $A_n \to 0$ when $c_s \to \Gamma$, rather than when $c_s \to 0$. With this model, the supply-constraint model of photosynthesis can be expressed as:

$$A_n = \frac{g_0}{1.6 \cdot (c_s - c_i) - g_1 \cdot h_s \cdot (c_s - \Gamma)} \tag{2.18}$$

Later, Leuning *et al.* (1995) made an additional modification to the model (Eq. 2.18) for C_3 plants as:

$$g_s = g_0 + \frac{a_1 \cdot A_n}{(c_s - \Gamma)\left(1 + \dfrac{D_s}{D_0}\right)} \tag{2.19}$$

where D_0 is the value of VPD at which stomatal conductance becomes zero. Note that D instead of VPD is used in Equation (2.19) and Equation (2.20).

Lloyd (1991) proposed that g_s is dependent on $\sqrt{D}$. Medlyn *et al.* (2011) further emphasized the importance of g_1 in the Ball-Berry model because of its sensitivity to environmental changes (*e.g.*, temperature, soil water and nutrients). They also agreed with Leuning *et al.* (1995) that VPD, instead of relative humidity, should be used in modeling $[A_n \sim g_s]$ for a new form of:

$$g_s = g_0 + 1.6 \cdot \left(1 + \frac{g_1}{\sqrt{D}}\right) \cdot \frac{A_n}{c_s} \tag{2.20}$$

In this model, g_1 is assumed to increase with the marginal water cost of carbon and with the CO_2 compensation point (Γ^*, kPa):

$$g_1 \propto \sqrt{\Gamma^* \cdot \lambda} \tag{2.21}$$

where λ (mol H_2O mol^{-1} CO_2) is a parameter, describing the marginal water cost of carbon gain (*i.e.*, similar meaning to water use efficiency).

There are four major empirical parameters in Ball-Berry model (Eq. 2.13): A_n, g_0, g_1, and h_s (or D). A demonstrative spreadsheet model is provided in S2-2 to show the role of these parameters in affecting stomatal conductance (Fig. 2.4). The curves of $[g_s \sim c_s]$ will be lowered with increases in g_1 values (Fig. 2.4a). Similarly, higher photosynthesis of a leaf is coupled with higher g_s and a faster decreasing trend (Fig. 2.4b). As for relative humidity, its absolute influences on g_s appears to be less than that of A_n and g_1 (Fig. 2.4c). Note that when h_s is replaced with D (Eqs. 2.19 and 2.20), water vapor effects are reversed; *i.e.*, a high D value means low relative humidity (*see* Chapter 1 for their relationships).

Independent estimates of g_s in abovementioned models becomes a necessity for predicting photosynthesis. Stomata opening is regulated by many environmental variables and depends on species, leaf age, time (phenophases), regional climate, disturbances, *etc.* Major hydro-meteorological influences include vapor pressure deficit (*i.e.*, reflecting both air temperature and absolute air moisture), level of radiation (*e.g.*, due to photoprohibition), and available water in the soil (*i.e.*, water supply). Jarvis (1976) proposed a

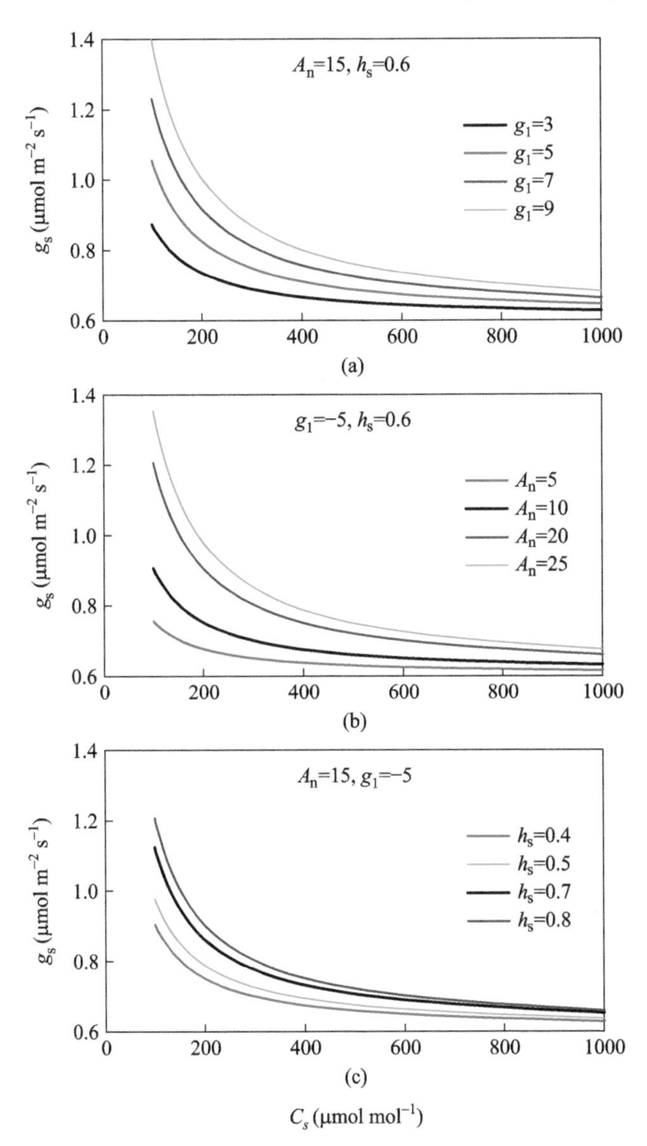

Fig. 2.4 Simulations of stomatal conductance (g_s) with different sets of parameters (Eq. 2.13). Other curves can be generated by altering parameters in S2-2 (Ball_Berry_model.xlsx).

VPD-driven g_s model:

$$g_s = g_{max} \cdot \left(1 - \frac{D}{D_0}\right) \tag{2.22}$$

From a soil water perspective, Jarvis (1976) also proposed a g_s model based on soil water potential (ψ, MPa):

$$g_s = g_{max} \cdot \left(1 - \frac{\psi}{\psi_0}\right) \tag{2.23}$$

where ψ_0 (MPa) is at which $g_s = 0$. Due to the fact that stomata can reduce their closure when incoming radiation (R) is high, Loustau *et al.* (1977) integrated R and VPD into a new empirical model:

$$g_s = g_{max} \cdot \frac{R_s}{k_1 + R_g} \cdot \frac{1 - k_2 \cdot D}{1 + k_3 \cdot D} \tag{2.24}$$

where R_s and R_g are the incoming solar radiation (W m^{-2}) and global radiation (W m^{-2}), respectively (*see* Chapter 1); and k_1 (W m^{-2}), k_2 (kPa^{-1}) and k_3 (kPa^{-1}) are empirical coefficients.

An obvious advantage of Ball-Berry-Leuning-Medlyn family models is that they allow us to model the exchange of CO_2 and H_2O, *i.e.*, coupling carbon and water cycles and changes over time (Dewar 2002). This is widely used in many ecosystem models for simulating carbon production and plant transpiration, including in the Community Land Model (Bonan *et al.* 2014). Meanwhile, these models involve numerous parameters that need to be measured or empirically estimated. While each of these parameters reflects specific physiological or physical process (*i.e.*, mechanistic foundations), their applications are more complicated and difficult than other models listed in Sections 2.2.1, 2.2.2 and 2.2.6.

2.2.5 Light Use Efficiency (LUE) Model

Ecosystem primary production (GPP, or NPP), or canopy photosynthesis (P_n), can be simply molded as a portion of PAR — light use efficiency (ε):

$$P_n = \varepsilon \cdot \text{PAR} \tag{2.25}$$

This model was proposed by Monteith (1972) for a tropical rain forest and tested for crop production in Britain (Monteith 1977). The LUE concept was theorized so that ecosystem production is proportional to the amount of energy (*i.e.*, absorbed PAR, or aPAR) used for photosynthesis that converts CO_2 to carbohydrate. A great challenge for applying this model is to estimate absorbed PAR that is directly affected by the amount of leaf in the canopy (*e.g.*, LAI), horizontal and vertical distributions of the leaves, foliar quality (*e.g.*, nitrogen content), species, soil and climatic conditions, and disturbances. Xiao *et al.* (2004) applied the following algorithm for aPAR:

$$\text{aPAR} = \text{fPAR} \cdot \text{PAR} \tag{2.26}$$

where fPAR is the fraction (0 – 1) of PAR that reaches leaves (or canopies), which can be estimated as a linear function of leaf area index (LAI) or vegetation index from remote sensing technology (*e.g.*, EVI) (Wu *et al.* 2010, John *et al.* 2013). Another challenge is that ε varies with climate, soil, vegetation, time of a day and day of year (Medlyn 1998). Nevertheless, an LUE-based model for estimating ecosystem primary production is simple, using PAR as the sole independent variable that is more available at ecosystem-regional-global scales. This advantage is the primary reason why the MODIS teams were able to measure global, continuous GPP based on Terra satellite data (Running *et al.* 2004). GPP is estimated as:

$$\text{GPP} = [\varepsilon_{\max} \cdot \text{mod}(\text{Temperature}) \cdot \text{mod}(\text{VPD})] \cdot \text{aPAR} \tag{2.27}$$

where $\varepsilon_{\max}$ is the maximum light use efficiency, and mod (Temperature) and mod (VPD) are modifiers (or scalars) to reduce $\varepsilon_{\max}$ under unfavorable temperatures and high VPD (Zhao *et al.* 2006). Other scalars can be added to the model. For example, a water-sensitive vegetation index (*i.e.*, land surface water index, LSWI) was used to calculate the relative value to the maximum potentials for a water scalar in the Vegetation Photosynthesis Model (VPM) (Xiao *et al.* 2004). While the LUE model mostly has been applied to estimating GPP, there is an increasing effort to apply the model for NPP. NPP can be estimated by modeling GPP first and then calculating NPP by assuming a portion of GPP is for NPP. Waring *et al.* (1998) suggested that the NPP/GPP ratio is conservatively (0.47±0.04). This ratio has since been revised to a range of 0.4 – 0.6 (Landsberg *et al.* 2020).

As demonstrated in Figure 2.5, there is a hypothesized "optimum" biophysical condition $\varepsilon_{\max}$ for a regulatory variable (*a.k.a.* modifier or scalar). Actual ε decreases from this optimum condition as stress increases. The rate of decrease can be modeled with linear, exponential, parabolic, and other forms, or set between known the minimum and maximum values (*i.e.*, the range). The range is used to normalize the scalar to (0, 1). It is worth noting that symmetric change of a regulatory variable from its optimum condition to the low/high extremes is widely practiced, albeit the reduction of ε from $\varepsilon_{\max}$ can be asymmetric. As an example, temperature scalar between minimum and optimum temperature can be different from that between the optimum and maximum. For estimating ε at a broader spatial scale (*i.e.*, landscape-region-globe), developing a reference table for each scalar is also an option (*e.g.*, Running *et al.* 2004). This re-scaling approach from the maximum values by including multiple environmental variables in all resource use models (Eq. 2.27) is one of the major philosophical approaches in ecological modeling (*e.g.*, the gap family models, and growth-yield models in forestry). For example, the gap class of models assume a geographical center has the optimum conditions for a species, whereas the distribution boundaries are used for the minimum value (Shugart 1984). This scalar-based modeling approach also can be applied to the other resource use models (Sections 2.2.7 – 2.2.9).

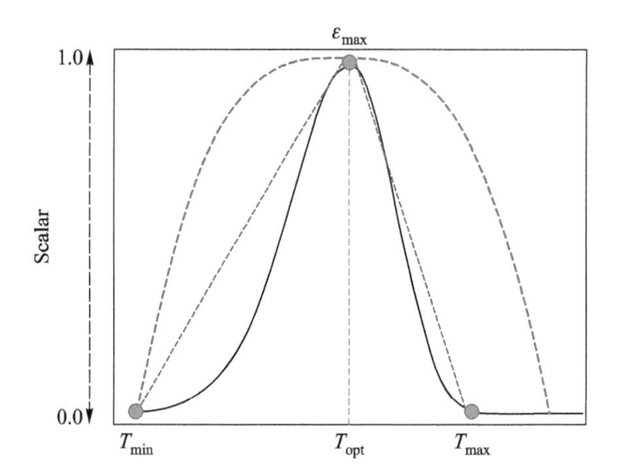

Fig. 2.5 Scalar development for modifying resource use efficiency (ε) from its maximum value (ε_{max}). Both symmetric and asymmetric functions can be used for estimating ε from ε_{max}. Minimum (T_{min}), maximum (T_{max}) and optimum (T_{opt}) temperature are used for deriving temperature scalar of three asymmetric approaches.

2.2.6 Nitrogen Use Efficiency (NUE) Model

Aber and Federer (1992) developed an ecosystem model — PnET — where they suggested that the maximum photosynthetic rate of an ecosystem (P_{max}) is a linear function of foliar nitrogen concentration, regardless of species and plant community structure. P_{max}, however, is modified by other biophysical variables such as temperature, available water, VPD, *etc.* P_{max} (μmol CO_2 m^{-2} s^{-1}) is calculated with a simple linear model based on a meta-analysis of prior publications:

$$P_{max} = \alpha + \beta \cdot N\% \tag{2.28}$$

where α and β are empirically estimated coefficients that value at -5.98 and 4.56, respectively, based on published data from 20 forests, and $N\%$ is the percentage of nitrogen in foliage of a stand (dry weight). Leaf gross photosynthesis (P_n) is calculated by assuming 10% of P_{max} is being used for basal respiration (*i.e.*, $P_n = 0.9 \times P_{max}$). P_n is further modified for suboptimal environmental conditions (*see* Section 2.2.6) as:

$$P_n = \alpha \cdot P_{max} \cdot \Delta T \cdot \Delta W \cdot \Delta \text{VPD} \tag{2.29}$$

where α is the portion of net photosynthesis in the gross photosynthesis (note the original PnET model assumes a value of 0.9); ΔT, ΔW, and ΔVPD are modifiers (or scalars) for temperature, available water, and vapor pressure deficit, respectively. All scalars have a value of 0–1 (*i.e.*, similar thinking in LUE, gap models). ΔT is calculated with a parabolic equation based on maximum and minimum daily P_{max}; ΔW is set as the mean level of water stress experienced in the previous month; and ΔVPD is calculated as VPD times a system specific constant (K). This equation is the core algorithm of the PnET model for simulating ecosystem NPP and ET at monthly step by treating an ecosystem as a big leaf, with a small number of model parameters, compared with other ecosystem models that use hundreds of parameters.

2.2.7 Water Use Efficiency (WUE) Model

Water use efficiency (WUE) has also been used for modeling ecosystem production (Tanner and Sinclair 1983). Briggs and Shantz (1913) introduced the concept of WUE by realizing that crop production needs support from sufficient water that can be improved through irrigation. Three major types of WUE have been practiced, including intrinsic WUE, instantaneous WUE, and ecosystem WUE. They are based CO_2 and H_2O exchanges through leaf stomata, gross photosynthesis and plant transpiration; gross primary productivity (parabolic); and actual evapotranspiration (ET_a), respectively. GPP has been substituted with NPP in some models.

Assuming CO_2 uptake and H_2O loss are coupled, GPP at ecosystem can be molded as:

$$\text{GPP} = \text{WUE} \cdot \text{ET} \tag{2.30}$$

An advantage of this model is predicting ecosystem production by monitoring ET loss, which is critical for intensively managed ecosystems such as crops (*i.e.*, through scheduling irrigation). Another advantage is to understanding the coupled carbon and water cycles under water stress (*e.g.*, drought), as well as under direct irrigation practices. Similar to the LUE models, challenges include empirical estimate of WUE and potential biophysical regulations for rescaling WUE (*e.g.*, Ito and Inatomi 2012).

2.2.8 Multiple Resource Use Efficiency (mRUE) Model

Multiple resources simultaneously regulate ecosystem production (Hodapp *et al.* 2019). Binkley *et al.* (2004) synthesized published results in a *Eucalyptus* plantation and concluded that the concept of RUE should at least include water, nutrients and light use. Han *et al.* (2016) integrated multiple resource use concepts in modeling GPP:

$$\begin{aligned}\text{GPP} &= \text{resource supply} \times \text{proportion of resource supply}\\ &\quad \times \text{captured efficiency of resource use}\end{aligned}$$

Here resource use efficiency for a specific resource (R_i) is defined as:

$$\text{RUE}_i = \frac{\text{GPP}}{R_i} \tag{2.31}$$

where R_i is the amount of absorbed resources, which can be expressed as available resource (R_{avail}) and an efficiency (ε_i) that ranges 0 – 1. When multiple RUEs are integrated, GPP can be modeled as:

$$\text{GPP} = (R_{\text{avail}_1} \times R_{\text{avail}_2} \times \cdots \times R_{\text{avail}_n})^{1/n} \cdot (\text{RUE}_1 \times \text{RUE}_2 \times \cdots \times \text{RUE}_n)^{1/n} \tag{2.32}$$

where n is the number of resource types to be considered. Multiple resource use (mRUE) is referred as $(\text{RUE}_1 \times \text{RUE}_2 \times \cdots \times \text{RUE}_n)^{1/n}$, and that of multiple ε_i (*i.e.*, $(\varepsilon_1 \times \varepsilon_2 \times \cdots \times \varepsilon_n)^{1/n}$) is referred as ε. Biophysical regulations of RUE and ε can be explored using approaches similar to LUE and NUE models (Sections 2.2.6 – 2.2.7). This model allows one to assess the regulation of all resources as an integrated subsystem, while their importance at different time periods or scales can be independently assessed. The interactions among various limiting resources can be examined and considered (Reed *et al.* 2020). Finally, resource use under constraints of temperature, moisture, disturbances, *etc.* can be included in the model.

2.3 The Datasets for Modeling Photosynthesis

The datasets (S2-3: Wang2018.xlsx) used for model demonstrations are from Wang *et al.* (2018). The authors measured net photosynthetic rate (A_n), stomatal conductance (g_s), and other relevant variables with a portable infrared gas analyzer (LI-COR 6400, LI-COR, Lincoln, Nebraska, USA) in an alpine meadow of the Tibetan Plateau (37°37′ N, 101°12′ E, 3240 a.s.l) during 2014 – 2016. The study was designed to assess how fertilizations of N and P may affect photosynthesis of two major plants: *Elymus dahuricus* and *Gentiana straminea.* Fully expanded healthy leaves were used in the experiment following the standard protocols of the LI-COR Company. In brief, leaves were exposed to a CO_2 concentration of 370 μmol mol^{-1}, at a leaf temperature of 25 °C, and with airflow through the chamber of 300 μmol s^{-1}. Leaves were acclimated to a photosynthetic photon flux density of 2000 μmol m^{-2} s^{-1} until photosynthetic rates stabilized. The rate of photosynthesis at a PPFD of 2000 μmol m^{-2} s^{-1} was defined as the net photosynthetic rate. Photosynthesis rates of different species and treatments were modeled with Farquhar's model (Eqs. 2.7, 2.11 and 2.12) and stomatal conductance was modeled with the Ball-Berry (Eq. 2.15) and Medlyn (Eq. 2.20) models (Fig. 2.6). Major model parameters (*e.g.*, V_{max}, J_{max}, WUE$_i$, K_c, R_d, *etc.*) are included in a supplementary Excel spreadsheet (S2-3: Wang2018.xlsx).

2.4 Model Performances

2.4.1 Light Response Models

Three versions of the Michaelis-Menten model (MM model) (Eqs. 2.2 – 2.4) and the Landsberg model (Eq. 2.5) are fitted for estimating leaf photosynthesis rate (P_n, μmol m^{-2} s^{-1}) for the two species used in Wang *et al.* (2018) (Fig. 2.7). These models performed well with a correlation coefficient of determination (r^2) of 0.67 for *E. dahuricus* and 0.52 for *G. straminea.* The three MM models produced almost identical results. More importantly, the MM model assumes the photosynthesis rate is zero only when PAR = 0 unless R_d (μmol m^{-2} s^{-1}) is included, whereas the Landsberg model has a light compensation point (I_{comp}). The Landsberg model also assumes a

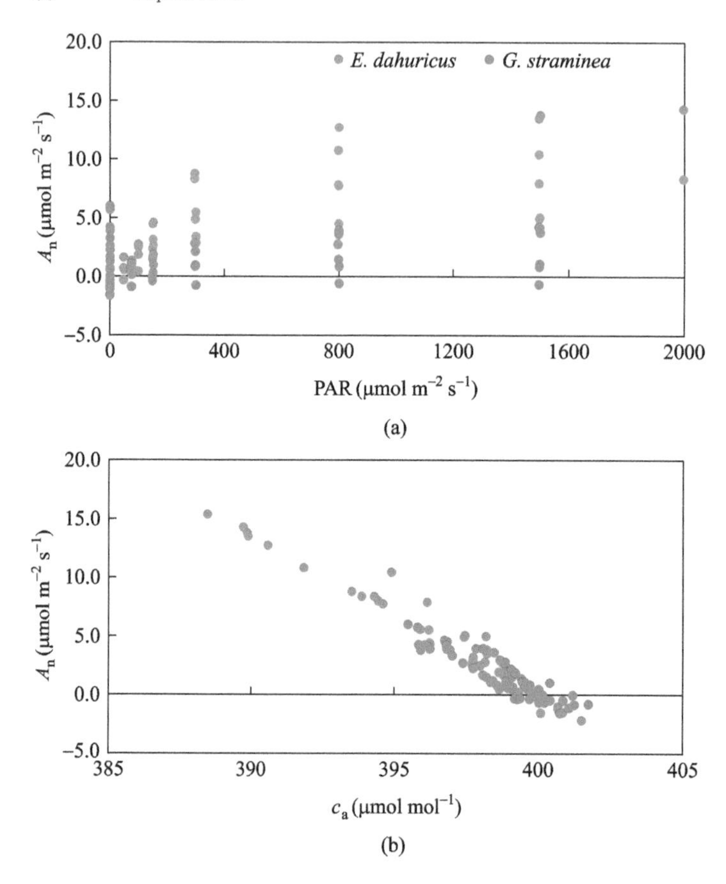

Fig. 2.6 Changes in photosynthesis rate (A_n) with (a) photosynthetically active radiation (PAR) and (b) CO_2 concentration (c_a) for two species in Wang *et al.* (2018) (data use permission received from the authors).

constant increasing rate of P_n with PAR (*i.e.*, α value); whereas the modified MM models (Eqs. 2.3 and 2.4) include a shape parameter (*i.e.*, β) to control the light response. Estimated model parameters, model comparisons, and model performances are included in the supplement spreadsheet LightR_models.xlsx (S2-4).

2.4.2 Results from Farquhar's Model

Photosynthesis rate of the two species studied in Wang *et al.* (2018) was modeled with three algorithms of Farquhar's model: A-c_i curve (Eq. 2.7), light-response curve (Eq. 2.11) and product-limited model (Eq. 2.12). Over

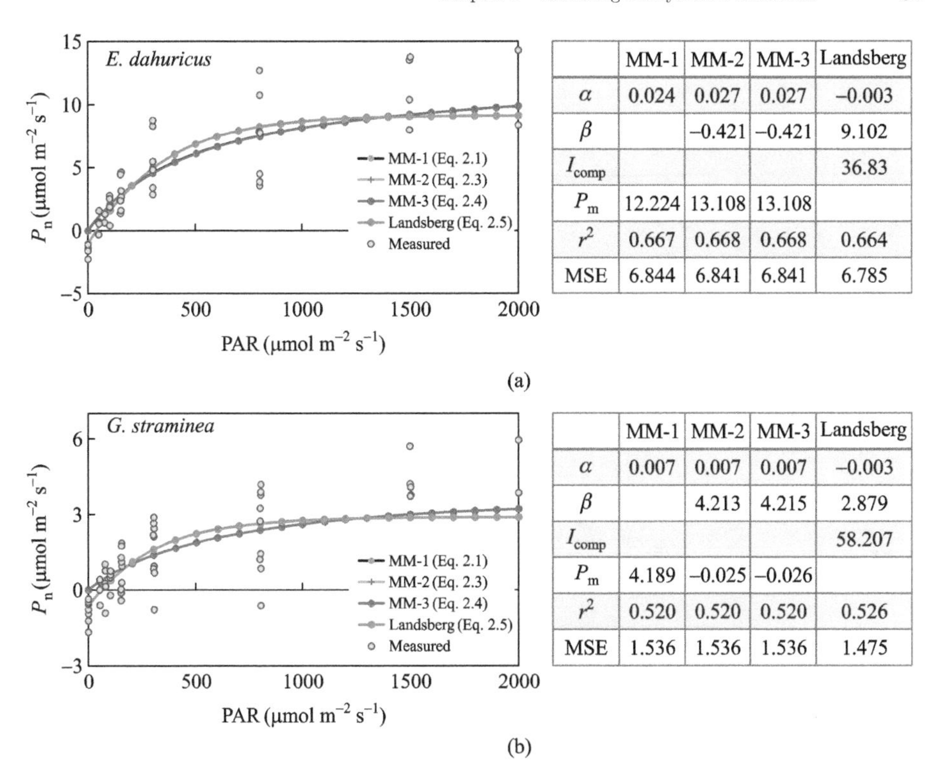

	MM-1	MM-2	MM-3	Landsberg
α	0.024	0.027	0.027	−0.003
β		−0.421	−0.421	9.102
I_{comp}				36.83
P_m	12.224	13.108	13.108	
r^2	0.667	0.668	0.668	0.664
MSE	6.844	6.841	6.841	6.785

	MM-1	MM-2	MM-3	Landsberg
α	0.007	0.007	0.007	−0.003
β		4.213	4.215	2.879
I_{comp}				58.207
P_m	4.189	−0.025	−0.026	
r^2	0.520	0.520	0.520	0.526
MSE	1.536	1.536	1.536	1.475

Fig. 2.7 Fitted light response curves using three Michaelis-Menten (MM) equations (Eqs. 2.2, 2.3 and 2.4) and the Landsberg model (Eq. 2.5) for two species on the Tibetan Plateau (Wang *et al.* 2018): (a) *E. dahuricus*, (b) *G. straminea*. Details are included in the supplement spreadsheet LightR_models.xlsx (S2-4).

the two-year study period, 33 leaves of each species were measured under four experimental treatments (*i.e.*, control, additional of N, P and N+P). The changes in photosynthesis rate with $V_{\max}$ (A_c model, Eq. 2.7; Fig. 2.8a) and $J_{\max}$ (A_j model, Eq. 2.11; Fig. 2.8b) showed a near identical $[A_n \sim J_{\max}]$ relationship between the two species. The $[A_n \sim V_{\max}]$ relationship appears more variable than $[A_n \sim J_{\max}]$, emphasizing the important roles of K and atmospheric vapor pressure (or VPD) in modeling A_c. Photosynthesis rate based on the rate of phosphate release (A_p model, Eq. 2.12) yields the highest among all 66 leaves. Surprisingly, the A_c model predicts overall higher values than the A_j model for both species, with 10 of 66 leaves having a lower photosynthesis rate based on the A_c model (Fig. 2.8c). This suggests that the photosynthesis rate of these two alpine species can be mostly modeled with the light response model (Eq. 2.11) or the light use efficiency model (Eq. 2.25).

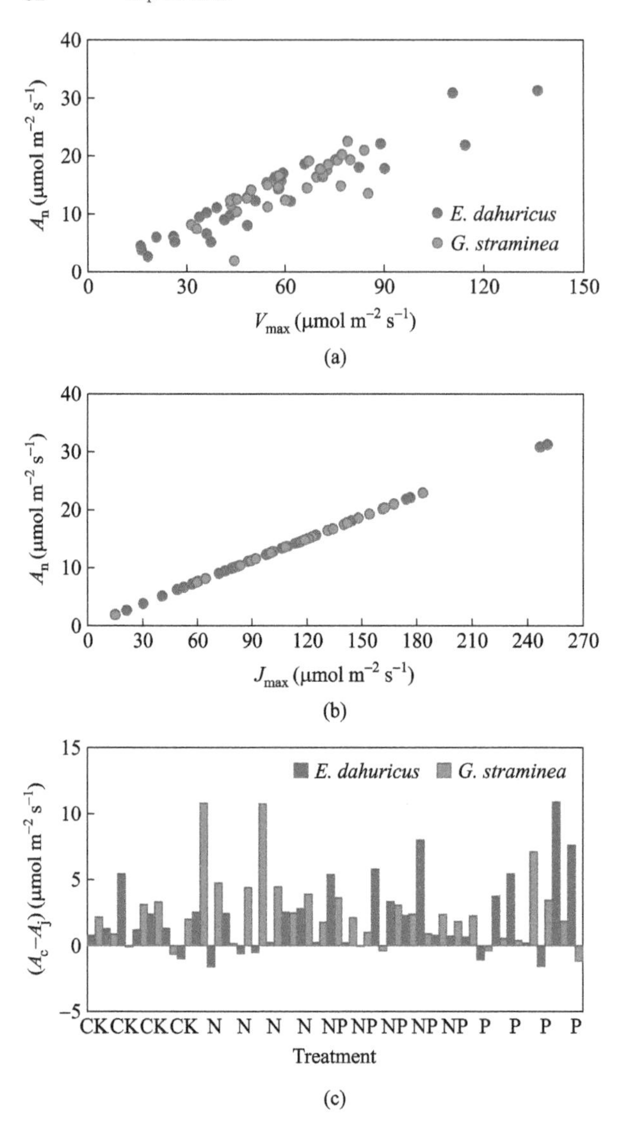

Fig. 2.8 Changes in photosynthesis rate (A_n) of two species in Wang *et al.* (2018) based on Farquhar's model (Eq. 2.6) with (a) the maximum rate of Rubisco (V_{max}) and (b) maximum rate of electron transport (J_{max}). Differences between Rubisco-limited model (Eq. 2.7) and light-limited model (Eq. 2.11) are shown in (c).

2.4.3 Results from Ball-Berry Model

Predicted stomatal conductance (g_s) of the two species using the Ball-Berry model (Eq. 2.13) is higher for *E. dahuricus* than for *G. straminea*, albeit their changes with photosynthesis rate (A_n) and the leaf surface CO_2 concentration (c_a) are similar (Fig. 2.9). The g_s value at a photosynthesis rate of zero (g_o; Eq. 2.15) is 0.0179 μmol m^{-2} s^{-1} for *E. dahuricus* and 0.0156 μmol m^{-2} s^{-1} for *G. straminea*. Because photosynthesis rate was mostly estimated with light-limited algorithm in Farquhar's model (Eq. 2.11, Fig. 2.8c), the $g_s \sim A_n$ relationship (Fig. 2.9a) reflects the regulatory role of the incoming radiation level, whereas the $g_s \sim c_a$ relationship accounts for the reduction in stomatal conductance under high CO_2 concentration (Fig. 2.9b).

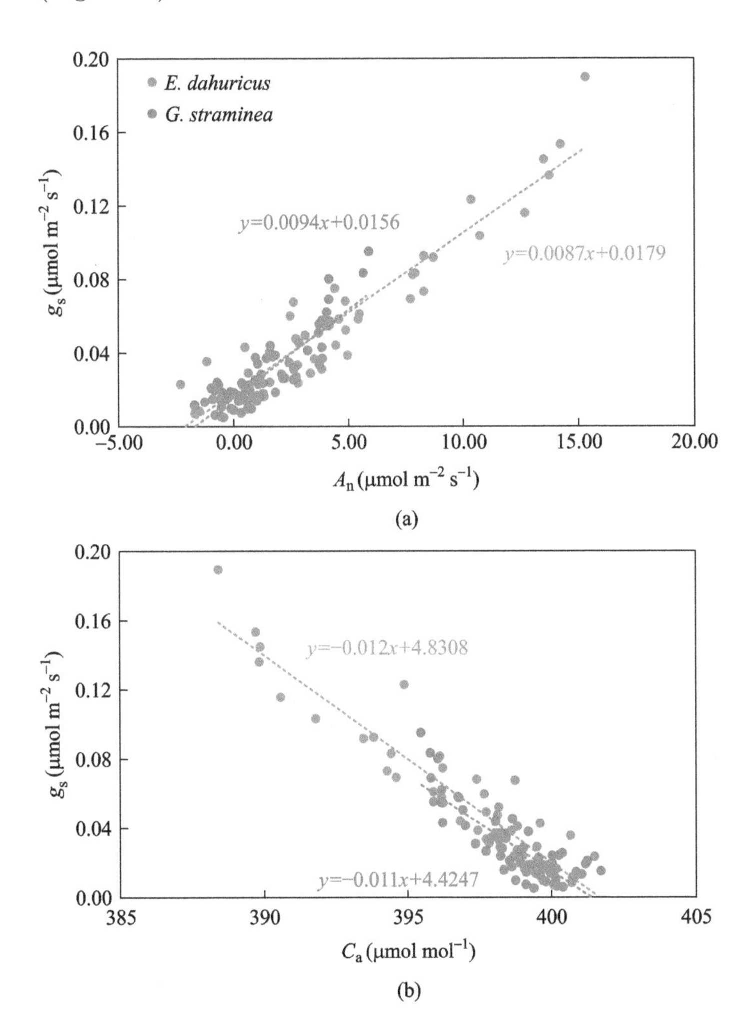

Fig. 2.9 Changes in stomatal conductance (g_s) with photosynthesis rate (A_n) and leaf surface CO_2 concentration for two species studied in Wang *et al.* (2018). A_n was estimated with Farquhar's model (Eq. 2.6) and g_s was estimated with the Ball-Berry model (Eq. 2.15). The data and regression results are included in the supplement document S2-3 (Wang2018.xlsx).

2.4.4 Other Models

Remote sensing modeling of GPP from various vegetation indexes (VIs), solar-induced fluorescence (SIF), and other land surface properties (*e.g.*, soil moisture, temperature) is emerging (*e.g.*, Gu *et al.* 2019). Performances of resource-based models (Eqs. 2.25 – 2.32) are not compared because these models are conventionally constructed through regression (linear or non-linear) analysis. They are applied at ecosystem-regional levels. The last four decades witnessed rapid growth in remote sensing technology, employing diverse platforms (satellite, airborne, Unmanned Aerial Vehicle or UAV) and sensors (passive and active). Remotely sensed data are particularly useful for estimating model parameters, such as leaf area index (LAI), vegetation index (VI), land surface temperature, soil moisture, canopy cover, phenology, *etc.* Various products of ecosystem structure and functions (*e.g.*, GPP, fractional cover, canopy structure) are estimated based on these indices (*e.g.*, GEDI LiDAR). Among the most relevant open data sources are NASA's Earthdata portal[1], Landsat products[2] since 1972, MOIDS products[3] since 2000 and Sentinel 1 and 2 of the European Space Agency[4] since 2014.

2.5 Summary

Empirical, mechanistic, and hybrid models have long been sought to describe and predict production of plants and terrestrial ecosystems. All models are based on assimilation through photosynthesis, which requires measurements of CO_2, H_2O, photosynthetically active radiation (PAR), and nutrients to address the underlying biological, ecological and physical processes. Here vegetation, soil and microclimatic conditions are used as regulatory factors for modeling the magnitude and changes of CO_2 assimilation. Plant physiologists and ecologists approached CO_2 assimilation from leaf and ecosystem perspectives, respectively, and developed their models more or less independently. Nevertheless, the approaches of modeling CO_2 assimilation from radiation (light response curve) and CO_2 concentration (A-c_i curve), along with biophysical regulations (*e.g.*, species vegetation characteristics, soil, and climate), have been the same in both schools of thought. With the

1 https://earthdata.nasa.gov/
2 https://earthexplorer.usgs.gov/
3 https://modis.gsfc.nasa.gov/data/dataprod/
4 https://sentinel.esa.int/web/sentinel/home

rapid development of computing technology and cross-disciplinary research, recent decades have witnessed fast integrations and merging of different models. Today, physiological models (*e.g.*, Farquhar's model, Ball-Berry model) are widely used in ecosystem models, including remote sensing modeling. Likewise, classical ecosystem production models (*e.g.*, the gap models) are increasingly considering detailed biophysical processes for model improvement.

Selections and uses of suitable models depend on study objectives and available resources (*e.g.*, data, infrastructure, duration, *etc.*). No single model can serve all purposes, but every model is suitable under certain conditions or specifications. With the emergence of Farquhar's photosynthesis model, and exponential increases in computing power, most ecosystem models predict GPP using light use efficiency approach scale Farquhar's model up in space and time with additional input parameters (*e.g.*, LAI, cover type, *etc.*). Resource use efficiency models are more commonly used at daily or longer time steps, while algorithms of Farquhar and Ball-Berry families appear more favored at hourly scale.

- Models based on light response curve are easy to understand and use. Only a few parameters $(2-4)$ are needed to construct these models. Much more effort is needed to examine the influences of other potential driving forces on model parameters.
- Physiological models have solid chemical and physical processes and theoretical foundations. Farquhar's model is based on the kinetic energy concept of the Michaelis-Menten model as well as the chemical processes of photosynthesis, whereas the Ball-Berry family of models are rooted in the gas diffusion process and the corresponding properties of gases and physical conditions.
- A large number of parameters $(5-10)$ are required for both Farquhar's model and the Ball-Berry model. These parameters are often difficult to measure or estimate. When these models are used to model ecosystem production, a tremendous amount of ancillary data on species composition, structure, soil conditions and microclimate are needed.
- Resource use models are also easy to understand and can be based on empirical parameters. They are particularly advantageous for modeling ecosystem production at landscape-region-global scales. These models have specific merits when applied with remote-sensed measures such as vegetation index, phenology, *etc.*

Online Supplementary Materials

S2-1: Light response curves through Michaelis-Menten and Landsberg models (LightResponse.xlsx).
S2-2: Simulations of stomatal conductance (g_s) based on the Ball-Berry model (Ball_Berry_model.xlsx).
S2-3: Field measurements and modeled photosynthesis rate (A_n, μmol m^{-2} s^{-1}) and parameters for two species in Wang *et al.* (2018) (Wang2018.xlsx).
S2-4: Model performances of Michaelis-Menten and Landsberg models for the two species in Wang *et al.* (2018) (LightR_models.xlsx).
S2-5: Python codes for estimating empirical coefficients through nonlinear regression analysis of Michaelis-Menten and Landsberg models (Chapter2_python.rar). This package has one dataset in Excel for practice and four Python programs for non-linear regression.

Scan the QR code or go to
https://msupress.org/supplement/BiophysicalModels
to access the supplementary materials.
User name: biophysical
Password: z4Y@sG3T

Acknowledgements

Several colleagues helped to form and improve this chapter. Richard Waring provided critical references and his insights on ecosystem production. Thomas Sharkey, Richard Waring, Ying-Ping Wang, and Dan Wang reviewed drafts of this chapter and provided very constructive suggestions and edits. Huimin Zou developed Python codes for the non-linear regression analysis of light response models. Dan Wang provided an impressive set of leaf-level measurements of photosynthesis for model demonstrations. She also gave many hours to explain the experiment, data collection, and their modeling processes. Vincenzo Giannico provided LiDAR image for Figure 2.2. Kristine Blakeslee edited the language and format of the draft.

References

Aber, J. D., and Federer, C. A. (1992). A generalized, lumped-parameter model of photosynthesis, evapotranspiration and net primary production in temperate and boreal forest ecosystems. *Oecologia*, 92(4), 463–474.

Ball, J. T., Woodrow, I. E., and Berry, J. A. (1987). A model predicting stomatal conductance and its contribution to the control of photosynthesis under different environmental conditions. In *Progress in Photosynthesis Research* (pp. 221–224). Springer, Dordrecht. http://doi.org/10.1007/978-94-017-0519-6-48.

Bellasio, C., Beerling, D. J., and Griffiths, H. (2016). An Excel tool for deriving key photosynthetic parameters from combined gas exchange and chlorophyll fluorescence: Theory and practice. *Plant, Cell & Environment*, 39(6), 1180–1197.

Bhuyan, U., Zang, C., Vicente-Serrano, S. M., and Menzel, A. (2017). Exploring relationships among tree-ring growth, climate variability, and seasonal leaf activity on varying timescales and spatial resolutions. *Remote Sensing*, 9(6), 526. http://doi.org/10.3390/rs 9060526.

Binkley, D., Stape, J. L., and Ryan, M. G. (2004). Thinking about efficiency of resource use in forests. *Forest Ecology and Management*, 193(1-2), 5–16.

Bonan, G. B., Williams, M., Fisher, R. A., and Oleson, K. W. (2014). Modeling stomatal conductance in the earth system: Linking leaf water-use efficiency and water transport along the soil-plant-atmosphere continuum. *Geoscientific Model Development*, 7(5), 2193–2222.

Botkin, D. B., Janak, J. F., and Wallis, J. R. (1972). Rationale, limitations, and assumptions of a northeastern forest growth simulator. *IBM Journal of Research and Development*, 16(2), 101–116.

Briggs, L. J., and Shantz, H. L. (1913). *The Water Requirement of Plants* (No. 284–285). US Government Printing Office.

Buckley, T. N. (2017). Modeling stomatal conductance. *Plant Physiology*, 174(2), 572–582.

Buckley, T. N., and Diaz-Espejo, A. (2015). Reporting estimates of maximum potential electron transport rate. *New Phytologist*, 205(1), 14–17.

Chapin III, F. S., Matson, P. A., and Vitousek, P. (2011). *Principles of Terrestrial Ecosystem Ecology*. Springer Science & Business Media. 529pp.

Chen, J., Brosofske, K. D., Noormets, A., Crow, T. R., Bresee, M. K., Le Moine, J. M., Euskirchen, E. S., Mather, S. V., and Zheng, D. (2004). A working framework for quantifying carbon sequestration in disturbed land mosaics. *Environmental Management*, 33(1), S210–S221.

Chen, J., Falk, M., Euskirchen, E., Paw U, K. T., Suchanek, T. H., Ustin, S. L., Bond, B. J., Brosofske, K. D., Phillips, N., and Bi, R. (2002). Biophysical controls of carbon flows in three successional Douglas-fir stands based on eddy-covariance measurements. *Tree Physiology*, 22(2-3), 169–177.

Chen, J., John, R., Sun, G., McNulty, S., Noormets, A., Xiao, J., Turner, M. G., and Franklin, J. F. (2014). Carbon fluxes and storage in forests and landscapes. In *Forest Landscapes and Global Change* (pp. 139–166). Springer, New York, 262pp.

Clutter, J. L., Fortson, J. C., Pienaar, L. V., Brister, G. H., and Bailey, R. L. (1983). *Timber Management: A Quantitative Approach.* John Wiley & Sons, Inc. 333 pp.

Dewar, R. C. (2002). The Ball-Berry-Leuning and Tardieu-Davies stomatal models: Synthesis and extension within a spatially aggregated picture of guard cell function. *Plant, Cell & Environment*, 25(11), 1383–1398.

Ehleringer, J., and Björkman, O. (1977). Quantum yields for CO_2 uptake in C_3 and C_4 plants: Dependence on temperature, CO_2, and O_2 concentration. *Plant Physiology*, 59(1), 86–90.

Ehleringer, J., and Cook, C. S. (1980). Measurements of photosynthesis in the field: Utility of the CO_2 depletion technique. *Plant, Cell & Environment*, 3(6), 479–482.

Evans, J. R. (1987). The dependence of quantum yield on wavelength and growth irradiance. *Functional Plant Biology*, 14(1), 69–79.

Evans, L. T. (1996). *Crop Evolution, Adaptation and Yield.* Cambridge University Press. 514pp.

Farquhar, G. D., Ehleringer, J. R., and Hubick, K. T. (1989). Carbon isotope discrimination and photosynthesis. *Annual Review of Plant Biology*, 40(1), 503–537.

Farquhar, G. D., von Caemmerer, S. V., and Berry, J. A. (1980). A biochemical model of photosynthetic CO_2 assimilation in leaves of C_3 species. *Planta*, 149(1), 78–90.

Farquhar, G. D., von Caemmerer, S. V., and Berry, J. A. (2001). Models of photosynthesis. *Plant Physiology*, 125(1), 42–45.

Franklin, J. F., Bledsoe, C. S., and Callahan, J. T. (1990). Contributions of the long-term ecological research program. *BioScience*, 40(7), 509–523.

Giannico, V., Lafortezza, R., John, R., Sanesi, G., Pesola, L., and Chen, J. (2016). Estimating stand volume and above-ground biomass of urban forests using LiDAR. *Remote Sensing*, 8(4), 339. http://doi.org/10.3390.rs8040339.

Gilmanov, T. G., Verma, S. B., Sims, P. L., Meyers, T. P., Bradford, J. A., Burba, G. G., and Suyker, A. E. (2003). Gross primary production and light response parameters of four Southern Plains ecosystems estimated using long-term CO_2-flux tower measurements. *Global Biogeochemical Cycles*, 17(2). https://doi.org/10.1029/2002GB002023.

Gu, L., Wood, J. D., Chang, C. Y., Sun, Y., and Riggs, J. S. (2019). Advancing terrestrial ecosystem science with a novel automated measurement system for sun-induced chlorophyll fluorescence for integration with eddy covariance flux networks. *Journal of Geophysical Research: Biogeosciences*, 124(1), 127–146.

Han, J., Chen, J., Miao, Y., and Wan, S. (2016). Multiple resource use efficiency (*mRUE*): A new concept for ecosystem production. *Scientific Reports*, 6, 37453.

Helliker, B. R., Song, X., Goulden, M. L., Clark, K., Bolstad, P., Munger, J. W., Chen, J., Noormets, A., Hollinger, D., Wofsy, S., and Martin, T. (2018). Assessing the interplay between canopy energy balance and photosynthesis with cellulose $\delta^{18}O$: Large-scale patterns and independent ground-truthing. *Oecologia*, 187(4), 995–1007.

Hodapp, D., Hillebrand, H., and Striebel, M. (2019). "Unifying" the concept of resource use efficiency in ecology. *Frontiers in Ecology and Evolution*, 6, 233. http://doi.org/10.3389/fevo.2018.00233.

Högberg, P., Nordgren, A., Buchmann, N., Taylor, A. F., Ekblad, A., Högberg, M. N., Nyberg, G., Ottosson-Löfvenius, M., and Read, D. J. (2001). Large-scale forest girdling shows that current photosynthesis drives soil respiration. *Nature*, 411(6839), 789–792.

Hollinger, D. Y., Kelliher, F. M., Byers, J. N., Hunt, J. E., McSeveny, T. M., and Weir, P. L. (1994). Carbon dioxide exchange between an undisturbed old-growth temperate forest and the atmosphere. *Ecology*, 75(1), 134–150.

Hunt, S. (2003). Measurements of photosynthesis and respiration in plants. *Physiologia Plantarum*, 117(3), 314–325.

Ito, A., and Inatomi, M. (2012). Water-use efficiency of the terrestrial biosphere: A model analysis focusing on interactions between the global carbon and water cycles. *Journal of Hydrometeorology*, 13(2), 681–694.

Jarvis, P. G. (1976). The interpretation of the variations in leaf water potential and stomatal conductance found in canopies in the field. *Philosophical Transactions of the Royal Society of London. B, Biological Sciences*, 273(927), 593–610.

John, R., Chen, J., Noormets, A., Xiao, X., Xu, J., Lu, N., and Chen, S. (2013). Modelling gross primary production in semi-arid Inner Mongolia using MODIS imagery and eddy covariance data. *International Journal of Remote Sensing*, 34(8), 2829–2857.

Kattge, J., and Knorr, W. (2007). Temperature acclimation in a biochemical model of photosynthesis: a reanalysis of data from 36 species. *Plant, Cell & Environment*, 30(9), 1176–1190.

Knapp, A. K., and Smith, M. D. (2001). Variation among biomes in temporal dynamics of aboveground primary production. *Science*, 291(5503), 481–484.

Kumarathunge, D. P., Medlyn, B. E., Drake, J. E., Rogers, A., and Tjoelker, M. G. (2019). No evidence for triose phosphate limitation of light-saturated leaf photosynthesis under current atmospheric CO_2 concentration. *Plant, Cell & Environment*, 42(12), 3241–3252.

Landsberg, J. J. (1977). Some useful equations for biological studies. *Experimental Agriculture*, 13(3), 273–286.

Landsberg, J. J., and Sands, P. (2011). *Physiological Ecology of Forest Production: Principles, Processes and Models* (Vol. 4). London: Elsevier/Academic Press. 352pp.

Landsberg, J. J., and Waring, R. H. (1997). A generalised model of forest productivity using simplified concepts of radiation-use efficiency, carbon balance and partitioning. *Forest Ecology and Management*, 95(3), 209–228.

Landsberg, J. J., Waring, R. H., and Williams, M. (2020). Commentary on the assessment of NPP/GPP ratio. *Tree Physiology*, 40(6), 695–699.

Lee, X., Massman, W., and Law, B. (2004). *Handbook of Micrometeorology: A Guide for Surface Flux Measurement and Analysis* (Vol. 29). Springer Science & Business Media. 250pp.

Leuning, R. (1990). Modelling stomatal behaviour and photosynthesis of *Eucalyptus grandis*. *Functional Plant Biology*, 17(2), 159–175.

Leuning, R. (1995). A critical appraisal of a combined stomatal-photosynthesis model for C_3 plants. *Plant, Cell & Environment*, 18(4), 339–355.

Lin, M., Wang, Z., He, L., Xu, K., Cheng, D., and Wang, G. (2015). Plant photosynthesis-irradiance curve responses to pollution show non-competitive inhibited Michaelis kinetics. *PloS One*, 10(11), e0142712. http://doi.org/10.1371/Journal.pone.0142712.

Lloyd, J. (1991). Modelling stomatal responses to environment in *Macadamia integrifolia*. *Functional Plant Biology*, 18(6), 649–660.

Ma, S., He, F., Tian, D., Zou, D., Yan, Z., Yang, Y., Zhou, T., Huang, K., Shen, H., and Fang, J. (2018). Variations and determinants of carbon content in plants: A global synthesis. *Biogeosciences*, 15(3), 693–702.

Medlyn, B. E. (1998). Physiological basis of the light use efficiency model. *Tree Physiology*, 18(3), 167–176.

Medlyn, B. E., Dreyer, E., Ellsworth, D., Forstreuter, M., Harley, P. C., Kirschbaum, M. U. F., Le Roux, X., Montpied, P., Strassemeyer, J., Walcroft, A., and Wang, K. (2002). Temperature response of parameters of a biochemically based model of photosynthesis. II. A review of experimental data. *Plant, Cell & Environment*, 25(9), 1167–1179.

Medlyn, B. E., Duursma, R. A., Eamus, D., Ellsworth, D. S., Prentice, I. C., Barton, C. V., Crous, K. Y., De Angelis, P., Freeman, M., and Wingate, L. (2011). Reconciling the optimal and empirical approaches to modelling stomatal conductance. *Global Change Biology*, 17(6), 2134–2144.

Michaelis, L., and Menten, M. L. (1913). Die Kinetik der Invertinwirkung. *Biochem Z*, 49: 333–369.

Monteith, J. L. (1972). Solar radiation and productivity in tropical ecosystems. *Journal of Applied Ecology*, 9(3), 747–766.

Monteith, J. L. (1977). Climate and the efficiency of crop production in Britain. *Philosophical Transactions of the Royal Society of London. B, Biological Sciences*, 281(980), 277–294.

Montesino-San Martín, M., Olesen, J. E., and Porter, J. R. (2014). A genotype, environment and management (G x E x M) analysis of adaptation in winter wheat to climate change in Denmark. *Agricultural and Forest Meteorology*, 187, 1–13.

Moualeu-Ngangue, D. P., Chen, T. W., and Stützel, H. (2017). A new method to estimate photosynthetic parameters through net assimilation rate-intercellular space CO_2 concentration (A-C_i) curve and chlorophyll fluorescence measurements. *New Phytologist*, 213(3), 1543–1554.

Muller, B., and Martre, P. (2019). Plant and crop simulation models: Powerful tools to link physiology, genetics, and phenomics. *Journal of Experimental Botany*, 70(9), 2339–2344.

Parton, W. J. (1996). The CENTURY model. In *Evaluation of Soil Organic Matter Models* (pp. 283–291). Springer, Berlin, Heidelberg. 433pp.

Pataki, D. E., Ehleringer, J. R., Flanagan, L. B., Yakir, D., Bowling, D. R., Still, C. J., Buchmann, N., Kaplan, J. Q., and Berry, J. A. (2003). The application and interpretation of Keeling plots in terrestrial carbon cycle research. *Global Biogeochemical Cycles*, 17(1): 1022.

Pearcy, R. W., Ehleringer, J. R., Mooney, H., and Rundel, P. W. (2012). *Plant Physiological Ecology: Field Methods and Instrumentation.* Springer Science & Business Media. 458pp.

Peaucelle, M., Bacour, C., Ciais, P., Vuichard, N., Kuppel, S., Peñuelas, J., Belelli Marchesini, L., Blanken, P. D., Buchmann, N., Chen, J., Delpierre, N., Desai, A. R., Dufrêne, E., Gianelle, D., Gimeno-Colera, C., Gruening, C., Helfter, C., Hörtnagl, L., Ibrom, A., Joffre, R., Kato, T., Kolb, T., Law, B., Lindroth, A., Mammarella, I., Merbold, L., Minerbi, S., Montagnani, L., Šigut, L., Sutton, M., Varlagin, A., Vesala, T.,

Wohlfahrt, G., Wolf, S., Yakir, D., and Viovy, N. (2019). Covariations between plant functional traits emerge from constraining parameterization of a terrestrial biosphere model. *Global Ecology and Biogeography*, 28(9), 1351–1365.

Reed, D. E., Chen, J., Abraha, M., Robertson, G. P., and Dahlin, K. M. (2020). The shifting role of mRUE for regulating ecosystem production. *Ecosystems*, 23(2), 359–369.

Reichstein, M., Falge, E., Baldocchi, D., Papale, D., Aubinet, M., Berbigier, P., Bernhofer, C., Buchmann, N., Gilmanov, T., Granier, A., Grünwald, T., Havrávanková, K., Ilvesniemi, H., Janous, D., Knohl, A., Laurila, T., Lohila, A., Loustau, D., Matteucci, G., Meyers, T., Miglietta, F., Ourcival, J-M., Pumpanen, J., Rambal, S., Rotenberg, E., Sanz, M., Tenhunen, J., Seufert, G., Vaccari, F., Vesala, T., Yakir, D., and Valentini, R. (2005). On the separation of net ecosystem exchange into assimilation and ecosystem respiration: Review and improved algorithm. *Global Change Biology*, 11(9), 1424–1439.

Running, S. W., Nemani, R. R., Heinsch, F. A., Zhao, M., Reeves, M., and Hashimoto, H. (2004). A continuous satellite-derived measure of global terrestrial primary production. *Bioscience*, 54(6), 547–560.

Schaefer, K., Schwalm, C. R., Williams, C., Arain, M. A., Barr, A., Chen, J. M., Davis, K. J., Dimitrov, D., Hilton, T. W., Hollinger, D.Y., Humphreys, E., Poulter, B., Raczka, B. M., Richardson, A. D., Sahoo, A., Thornton, P., Vargas, R., Verbeeck, H., Anderson, R., Baker, I., Black, T. A., Bolstad, P., Chen, J., Curtis, P., Desai, A. R., Dietze, M., Dragoni, D., Gough, C., Grant, R. F., Gu, L., Jain, A., Kucharik, C., Law, B., Liu, S., Lokipitiya, E., Margolis, H. A., Matamala, R., McCaughey, J. H., Monson, R., Munger, J. W., Oechel, W., Peng, C., Price, D. T., Ricciuto, D., Riley, W. J., Roulet, N., Tian, H., Tonitto, C., Torn, M., Weng, E., and Zhou, X. (2012). A model-data comparison of gross primary productivity: Results from the North American Carbon Program site synthesis. *Journal of Geophysical Research: Biogeosciences*, 117(G3). http://doi.org/10.1029/2012JG001960.

Schlesinger, W. H. (1977). Carbon balance in terrestrial detritus. *Annual Review of Ecology and Systematics*, 8(1): 51–81.

Sellers, P. J., Mintz, Y. C. S. Y., Sud, Y. E. A., and Dalcher, A. (1986). A simple biosphere model (SiB) for use within general circulation models. *Journal of the Atmospheric Sciences*, 43(6), 505–531.

Sharkey, T. D. (2019). Is triose phosphate utilization important for understanding photosynthesis? *Journal of Experimental Botany*, 70(20), 5521–5525.

Sharkey, T. D. (2015). What gas exchange data can tell us about photosynthesis. *Plant, Cell & Environment*, 39(6), 1161–1163.

Sharkey, T. D., Bernacchi, C. J., Farquhar, G. D., and Singsaas, E. L. (2007). Fitting photosynthetic carbon dioxide response curves for C_3 leaves. *Plant, Cell & Environment*, 30(9), 1035–1040.

Shugart, H. H. (1984). *A Theory of Forest Dynamics: The Ecological Implications of Forest Succession Models*. Springer-Verlag. 278pp.

Tanner, C. B., and Sinclair, T. R. (1983). Efficient water use in crop production: Research or re-search? In *Limitations to Efficient Water Use in Crop Production* (pp1–2). Madison, WI: American Society of Agronomy.

Thornley, J. H. (2011). Plant growth and respiration re-visited: Maintenance respiration defined—it is an emergent property of, not a separate process within, the system — and why the respiration: Photosynthesis ratio is conservative. *Annals of Botany*, 108(7), 1365–1380.

Thornley, J. H. M., and Johnson, I. R. (1990). *Plant and Crop Modelling —A Mathematical Approach to Plant and Crop Physiology.* Clarendon Press, Oxford, 660 pp.

Von Caemmerer, S. (2013). Steady-state models of photosynthesis. *Plant, Cell & Environment*, 36(9), 1617–1630.

Wang, D., Ling, T., Wang, P., Jing, P., Fan, J., Wang, H., and Zhang, Y. (2018). Effects of 8-year nitrogen and phosphorus treatments on the ecophysiological traits of two key species on Tibetan Plateau. *Frontiers in Plant Science*, 9, 1290.

Waring, R. H., Landsberg, J. J., and Williams, M. (1998). Net primary production of forests: A constant fraction of gross primary production? *Tree Physiology*, 18(2), 129–134.

Waring, R. H., and Running, S. W. (2010). *Forest Ecosystems: Analysis at Multiple Scales.* Elsevier. 440pp.

Wolf, A., Akshalov, K., Saliendra, N., Johnson, D. A., and Laca, E. A. (2006). Inverse estimation of V_{cmax}, leaf area index, and the Ball-Berry parameter from carbon and energy fluxes. *Journal of Geophysical Research: Atmospheres*, 111(D8). https://doi.org/10.1029/2005JD005927.

Wu, C., Munger, J. W., Niu, Z., and Kuang, D. (2010). Comparison of multiple models for estimating gross primary production using MODIS and eddy covariance data in Harvard Forest. *Remote Sensing of Environment*, 114(12), 2925–2939.

Xiao, X., Zhang, Q., Braswell, B., Urbanski, S., Boles, S., Wofsy, S., Moore III, B., and Ojima, D. (2004). Modeling gross primary production of temperate deciduous broadleaf forest using satellite images and climate data. *Remote Sensing of Environment*, 91(2), 256–270.

Yakir, D., and Sternberg, L. da S. L. (2000). The use of stable isotopes to study ecosystem gas exchange. *Oecologia*, 123(3), 297–311.

Zhao, M., Running, S. W., and Nemani, R. R. (2006). Sensitivity of Moderate Resolution Imaging Spectroradiometer (MODIS) terrestrial primary production to the accuracy of meteorological reanalyses. *Journal of Geophysical Research: Biogeosciences*, 111(G1). doi:10.1029/2004JG000004.

Chapter 3
Modeling Ecosystem Respiration

Jiquan Chen

3.1 Introduction

Ecosystem respiration is broadly referred to as the carbon dioxide (CO_2) released from an ecosystem to the atmosphere. It may comprised of autotrophic (R_a) from living plant components and heterotrophic (R_h) respiration due to decomposition of organic matter (Chapter 2). In some instances, CO_2 may also be released from weathering of bedrock (Chapter 2). Ecosystem respiration is the second largest carbon (C) flux in most terrestrial ecosystems after photosynthetic uptake. Some ecosystems experience CO_2 flux, as well, for example wetlands, old-growth forests, or ecosystems that are experiencing extreme events (*e.g.*, drought) or natural or human induced disturbances (*e.g.*, fire, insect defoliation, harvesting or grazing of grasslands). It includes terms of above- and below-ground autotrophic (R_a) and heterotrophic (R_h) respiration, fauna respiration and CO_2 release from weathering of bedrock. The last two terms, as well as heterotrophic respiration of aboveground live biomass, are small and negligible in most terrestrial ecosystems (Harmon *et al.* 2011, Chen *et al.* 2014). The sum of belowground components of R_a and R_h is conventionally referred as soil respiration (R_s). The respiration is often expressed as μmol CO_2 m^{-2} s^{-1}, or

Jiquan Chen
Landscape Ecology & Ecosystem Science (LEES) Lab, Department of Geography, Environment, and Spatial Sciences & Center for Global Change and Earth Observations, Michigan State University, East Lansing, MI 48823
Email: jqchen@msu.edu

Jiquan Chen, *Biophysical Models and Applications in Ecosystem Analysis*,
https://doi.org/10.3868/978-7-04-055256-0-3

mg C m^{-2} s^{-1}, where the molar mass of CO_2 is 44 g mol^{-1}, with 27.29% carbon or 12 g mol^{-1}. Depending on the measurements and study objectives, ecosystem respiration is reported at hourly, daily, or yearly time scale. Aboveground autotrophic respiration in arid and semi-arid ecosystems is small, while aboveground heterotrophic respiration in rain forests and old-growth forests potentially can be high. Most respiration studies focus on belowground respiration, with sporadic efforts to examine respirations from canopies, tree trunks, and understory vegetation (*see* Law *et al.* 1999, Lavigne *et al.* 1996, 2003, Gough *et al.* 2007, Li *et al.* 2012).

Scientific investigations of respiration can be traced to the late 1800s, when German scholars measured CO_2 concentration in soils (Peterson 1870), which were included in agricultural handbooks in 1878 (Wollny 1880, 1881). In the early 1900s, carbon production in soils was already considered an indicator of soil productivity (*e.g.*, Neller 1918). The earliest descriptions of respiration were likely from Leather (1915) and Lundegårdh (1922), where soil respiration was described as "the amount of CO_2 produced by a certain soil area during a certain time" (Lieth and Ouellette 1962). Endeavors to examine soil respiration have been promoted since the late 1960s, when the International Geosphere-Biosphere Programme (IGBP) initiated global studies on ecosystem productivity of different biomes, where understanding of the respiratory losses of carbon was a major focus (*e.g.*, Schlesinger 1977, Raich and Nadelhoffer 1989).

Measuring CO_2 in natural systems can be traced to the mid-1900s (Pettenkofer 1861). Before the 1870s measurements of respiration were mostly based on incubations of samples (*e.g.*, soil, or organic materials) (*e.g.*, Lundegårdh 1927), but these have been replaced with *in-situ* chamber-based or micrometeorological approaches since the 1990s (Falge *et al.* 2002, Ryan and Law 2005, Harmon *et al.* 2011). The devolvement of automated chambers (Fig. 3.1) promoted studies beyond soils and included specific components ecosystems, such as leaves, stems, litter, and down logs in a forest (*e.g.*, Li *et al.* 2012). Field measurement studies also include manipulative experiments (*e.g.*, Euskirchen *et al.* 2003, Concilio *et al.* 2005, DeForest *et al.* 2006) or focus on the changes in ecosystem composition and structure (*e.g.*, Li *et al.* 2012). Since 1990s, there has been an increass in studies that are based on measurements from eddy-covariance flux towers (*see* Fig. 2.2b) because these measure the net ecosystem exchange (NEE) of CO_2 between an ecosystem and the atmosphere because measurements taken in dark (*e.g.*, nighttime, or chambers under cover) are respiration. Nighttime data from eddy-covariance towers (Fig. 3.2b) is considered ecosystem respiration (Reichstein *et al.* 2005, Gorsel *et al.* 2009). Chamber-based approaches are relatively inexpensive and suitable for small areas (less than a few square meters), whereas tower-based measurements are costly and difficult to obtain but can provide an integrated measure of an ecosystems at several hectres to hundreds hectres of square meters (Fig. 2.2b). Additionally, carbon and oxygen isotopic compositions have been used to tease apart different respiration components (Yakir and Sternberg 2000).

Fig. 3.1 One of the 24 automated chambers at the Mt. Fuji Flux Site in central Japan.

Unlike photosynthesis (Chapter 2) and evapotranspiration (Chapter 4) that have mechanistic bases (*e.g.*, Michaelis-Menten kinetics), soil and ecosystem respiration is modeled empirically. These models require direct measurement of respiration as well as the independent variables for model validation, algorithms establishment, and estimates of empirical coefficients through regression analysis. The most dominant driving force on respiration is temperature of the soil or air. In general, respirations of soil, dead and live organic materials, or ecosystems increases with temperature. The theoretical foundation of the relationship between respiration and temperature was initially proposed by a Dutch chemist Jacobus Henricus Van't Hoff in 1884 (Van't Hoff 1884). His concept was that the change in the equilibrium constant of a chemical reaction to the change in temperature drives the standard enthalpy change for a process. For example, change in respiration with temperature can be expressed as a Q_{10} model (Van't Hoff 1898):

$$Q_{10} = \left(\frac{R_2}{R_1}\right)^{\left(\frac{10}{T_2 - T_1}\right)} \tag{3.1}$$

where respiration rate is measured as R_1 under temperature T_1 and R_2 is measured at temperature T_2. Q_{10} (a unitless measure) describes the reaction rate increase when the temperature is raised by 10 °C (or K). This model

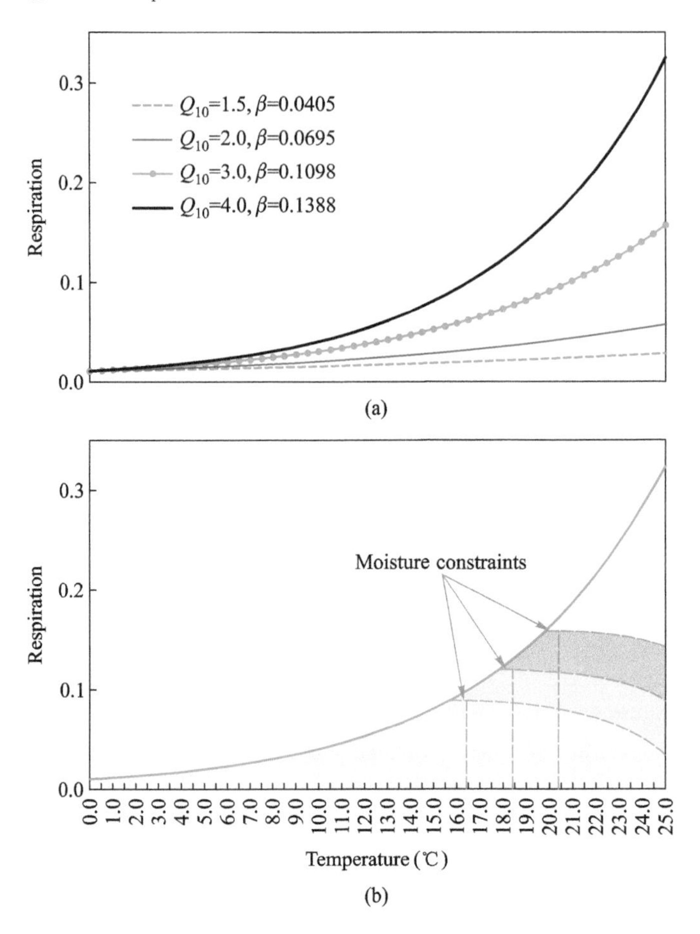

Fig. 3.2 (a) Schematic illustration of change in respiration with temperature by an exponential function (Eq. 3.3) for four Q_{10} values; (b) The exponential increase of respiration can be limited by other ecological resources such as moisture. The respiration reduction due to low moisture can be linear, polynomial, Gamma, logistic, or other forms. The threshold point can be empirically determined for a site or a specific time period.

(Eq. 3.1) is often used in the literature and has been expressed as well:

$$R = R_0 \cdot Q_{10}^{\frac{(T_2 - T_1)}{10}} \tag{3.2}$$

where R_0 is called reference respiration at 0 °C. The change in R with temperature can be expressed as linear, exponential, or in other complex forms (*see* Section 3.2). An exponential form is also widely used in respiration studies as:

$$R = \alpha \cdot \mathrm{e}^{\beta \cdot T} \tag{3.3}$$

where β is the rate of change with increasing temperature and α is the respiration at near zero temperature (°C). Q_{10} is calculated as:

$$Q_{10} = \mathrm{e}^{10 \cdot T} \tag{3.4}$$

The introduction of the Q_{10} concept allowed comparing respiration among samples, such as among different ecosystems or in one ecosystem in different time periods, or under different climatic conditions and disturbance regimes. Across the global terrestrial ecosystem, Q_{10} varies from 1 to 10, depending on the geographic location and ecosystems characteristics (Xu and Qi 2001). Davidson *et al.* (1998) reported a range of 3.4 to 5.6 for soil respiration among several temperate mixed-hardwood forests. A synthesis of global soil respiration by Raich and Schlesinger (1992) indicated a Q_{10} value of 1.3 to 3.3, with an overall mean of 2.4. Ecosystem respiration analysis conducted by Mahecha *et al.* (2010) based on the global FLUXNET dataset that included observations from 60 flux tower sites showed a Q_{10} value of 1.4 ± 0.1, which is insensitive of mean annual air temperature and biome. Q_{10} models were developed by scientists mostly for cool, temperate climates. It might not be useful for some ecosystems (*e.g.*, desert, boreal peatlands) on a global scale because of complications from other driving forces (*e.g.*, water table in wetlands, extreme low soil moisture in arid regions).

Ecosystem respiration, or its components such as soil respiration, is the result of complex interactions among biological, ecological and physical processes. Increasing evidence shows that simulations of ecosystem respiration needs to include more than just soil or air temperature. In modeling soil respiration, soil moisture has been proposed as another significant independent variable, because available water is crucial for many biological and physical processes (*e.g.*, Davidson *et al.* 2006, Curiel *et al.* 2007, Giasson *et al.* 2013). For example, Meyer *et al.* (2018) concluded that amount of soil organic carbon (SOC) or belowground biomass, carbon quality, pH level, nitrogen content, C : N ratio, mean annual precipitation (MAP), and mean annual temperature (MAT) are needed in modeling soil respiration. Other studies show that including snow cover (Wang *et al.* 2013), disturbances (*e.g.*, fertilization, irrigation), climatic extremes, plant community composition (*e.g.*, C_3 *vs.* C_4 plants) and structure, phenology, vigor of plant growth (*i.e.*, photosynthesis; Högberg *et al.* 2001), and timing during a day or day of year may help in accurate modeling of soil or ecosystem respiration (*see* DeForest *et al.* 2006, Khomik *et al.* 2006, Richardson *et al.* 2006, Xu *et al.* 2011).

3.2 Models for Ecosystem Respiration

Over a dozen model forms have been proposed and applied in the literature. These models are empirical and have to be constructed using *in-situ* measurements of respiration. Some models can be used to calculate Q_{10} values based on Van't Hoff's principle. Luo and Zhou (2010) presented 11 different models, whereas Richardson *et al.* (2006) expanded the time dimension in 12 types of respiration models. These models cover simple linear, log-linear (Eq. 3.3), Gamma function (Khomik *et al.* 2009), or sophisticated time series (Richardson *et al.* 2006). Some models included other ecosystem/ soil properties, such as soil moisture, day of year, litter depth, and multiple temperature measurements, as additional independent variables. Modeling respiration based on complex relationships also has been attempted (*e.g.*, Richardson *et al.* 2006). An overview of the respiration models is presented below. They are mostly applied for soil respiration, but they can be applied for modeling respiration of an ecosystem or its components.

3.2.1 Linear and Log-linear Models

These are simple linear models for predicting respiration (R) using temperature (T, $^{\circ}$C):

$$R = \alpha + \beta \cdot T \tag{3.5}$$

where α and β represent the basal respiration at near zero temperature (R_0, $^{\circ}$C) and rate of R increase per degree (*i.e.*, a constant), respectively. The Q_{10} value is calculated as ($10 * \beta$). Although this model has not been widely applied because of large error terms (*see* Section 3.4, Table. 3.1), its simplicity and easy use have some merit in modeling respiration, especially when R varies substantially for a given temperature. To improve model precision, a natural logarithm linear model (*i.e.*, Eq. 3.3) has been applied instead:

$$\ln(R) = \ln(\alpha) + \beta \cdot T \tag{3.6}$$

This model assumes that the changing rate of respiration with temperature is an exponential function of temperature. An obvious pitfall of both linear models is that empirically estimated α and β cannot be compared among different respiration terms, among ecosystems, or in different times, because α and β are correlated. Replacing α value with the basal respiration (R_0) at $T = 0\ ^{\circ}$C will partially resolve the problems but reduce model's accuracy.

3.2.2 Quadratic and Polynomial Models

Both quadratic and polynomial models have been used to predict respiration with high level of confidence (Wofsy 1993, Yu *et al.* 2011). The quadratic model is expressed as:

$$R = \alpha + \beta_0 \cdot T + \beta_1 \cdot T^2 \tag{3.7}$$

The quadratic model assumes that the rate of respiration change with temperature is a linear function of T and is symmetric around an optimum temperature, which is one of the mathematical properties. For site-specific modeling, a polynomial equation of various orders can also be applied, such as (Wofsy *et al.* 1993):

$$R = \alpha + \beta_1 \cdot T + \beta_1 \cdot T^2 + \beta_3 \cdot T^3 + \beta_4 \cdot T^4 + \beta_5 \cdot T^5 \tag{3.8}$$

The polynomial equation can provide accurate predictions but lacks any theoretical foundation and should not be used beyond the range of *in-situ* measurements.

3.2.3 Arrhenius Model

The Arrhenius form of the model proposed by Lloyd and Taylor (1994) has been widely used as the algorism in modeling respiration. There exist several forms of the model as well (*a.k.a.*, Lloyd-Taylor model). Görres *et al.* (2016) presented the model as:

$$R = R_{10} \cdot \mathrm{e}^{E_0\left[\frac{1}{56.02} - \frac{1}{T-227.13}\right]} \tag{3.9}$$

where R_{10} is the respiration rate at a reference temperature of 10 °C (*a.k.a.* reference respiration), E_0 is the temperature sensitivity coefficient (K), and T is soil temperature at a certain (*e.g.*, 5 cm) depth (K). Temperature in Kelvin units is used, which is the base unit of thermodynamic temperature measurement in the International System of Units (SI) of measurement

(K = °C + 273.15). Introduction of R_{10} allows comparisons among ecosystems or at different times, by avoiding the correlation between the two coefficients in Equations (3.5)–(3.6), albeit the confidence level (*e.g.*, correlation coefficient of determination, or r^2) is also lowered. The reference respiration is recommended to be site and time specific (Yuan *et al.* 2011, Perkins *et al.* 2012). Other types of expressions of the same form can be found in Reichstein *et al.* (2005), Richardson and Hollinger (2005), Perkins *et al.* (2012), Hill *et al.* (2018) and others.

3.2.4 Logistic Model

Barr *et al.* (2002) revised the classical logistic model to a simple form to model respiration:

$$R = \frac{\alpha}{1 + e^{(\beta_0 - \beta_1 \cdot T)}} \tag{3.10}$$

where α, β_0, and β_1 are estimated coefficients and T is temperature in °C. The logistic model assumes that the rate of change in respiration with temperature is not a constant, but peaks at a specific temperature and eventually returns to zero at high temperature. The precise temperature at 1/2 of the maximum R can be calculated.

3.2.5 Gamma Model

Khomik *et al.* (2009) proposed an approach, based on integrals of the Gamma function, for predicting respiration because prevalent respiration models in the literature had limitations when they were applied across a wide range of ecosystems or climates. They also did not allow R to decrease at high temperatures when respiration was constrained (*e.g.*, Fig. 3.2b). Gamma model helped to address these aspects. Gamma model is expressed as:

$$R = T^{\alpha} \cdot e^{\beta_0 + \beta_1 \cdot T} \tag{3.11}$$

where α, β_0 and β_1 are empirical coefficients to be estimated with measurements. Khomik *et al.* (2009) stated that this model has two strong features: exponentiality and power. When α is equal to zero, the model becomes an exponential function, and when β_1 is equal to zero, the model becomes a power function. Furthermore, unlike quadratic functions that would also allow R to decrease at high T values, Gamma model provides asymmetric changes by its maximum value. The T (°C) value at which R peaks (*i.e.*, $T_{\max}$, °C) can be determined as:

$$T_{\max} = \frac{\alpha}{-\beta_1} - 40 \tag{3.12}$$

For convenient estimation of the coefficients, the model can be linearized as:

$$\ln(R) = \alpha \cdot \ln(T) + \beta_0 + \beta_1 \cdot T \tag{3.13}$$

From Equation (3.11), it appears that this model assumes that the logarithm of respiration is a linear combination of logarithm and linear function of temperature, which is similar to the model of DeForest *et al.* (2006) (Eq. 3.14). Khomik *et al.* (2009) point out that an additional advantage of the linear form of Gamma model is that it can incorporate additional explanatory variables (into Eq. 3.13) as qualitative or quantitative values (*e.g.*, moisture, root biomass, *etc.*).

3.2.6 Biophysically Constrained Models

Respirations from different components of an ecosystem are simultaneously modulated by other biophysical conditions (*i.e.*, covariates). Soil respiration is significantly regulated by soil moisture, nutrients, root biomass, aboveground photosynthesis, *etc.* In some cases, these variables are better predictors of respiration (*see* example in Xu *et al.* 2011). For example, Euskirchen *et al.* (2003) reported that litter depth contributed more to the variations in soil respiration than soil temperature among the six temperate forest ecosystems. Inclusions of these variables in respiration modeling is highly recommended. For drylands and Mediterranean ecosystems, soil water content can suppress respiration when its value decreases to the threshold point (Xu and Qi 2001). Similarly, submerged wetlands (*i.e.*, high water table) have very different respiration-temperature relationships.

Numerous moisture-included respiration models exist, with most of them focusing on soil respiration (*e.g.* Ma *et al.* 2004, DeForest *et al.* 2006). A simple approach is to assume that respiration is linearly, or quadratically, suppressed by availability of water in soil (*see* Fig. 3.3b). In many terrestrial ecosystems, soil moisture is often negatively related to soil temperature (*e.g.*, Fig. 3.3a). This results in a reduced respiration rate after a temperature threshold. When water was available, water was a poor predictor of soil respiration; only when it was limiting did it make a difference in modeling respiration. In other words, it is the lack of water, not abundance, that mediates soil respiration. It is worth noting that available water is a function of soil texture although it is rarely reported as water potential rather than water content. DeForest *et al.* (2006) assumed that this reduction is a linear function of soil moisture (θ). Soil respiration based on the exponential model (Eq. 3.3) was tried as:

$$R = \left(R_{10} \cdot \mathrm{e}^{\beta \cdot T}\right) + (a \cdot \theta + b) \tag{3.14}$$

where a and b were empirically estimated from the residuals after the exponential term with θ. This model was successfully modified for modeling respiration of soil-to-snow profiles (Contosta *et al.* 2016).

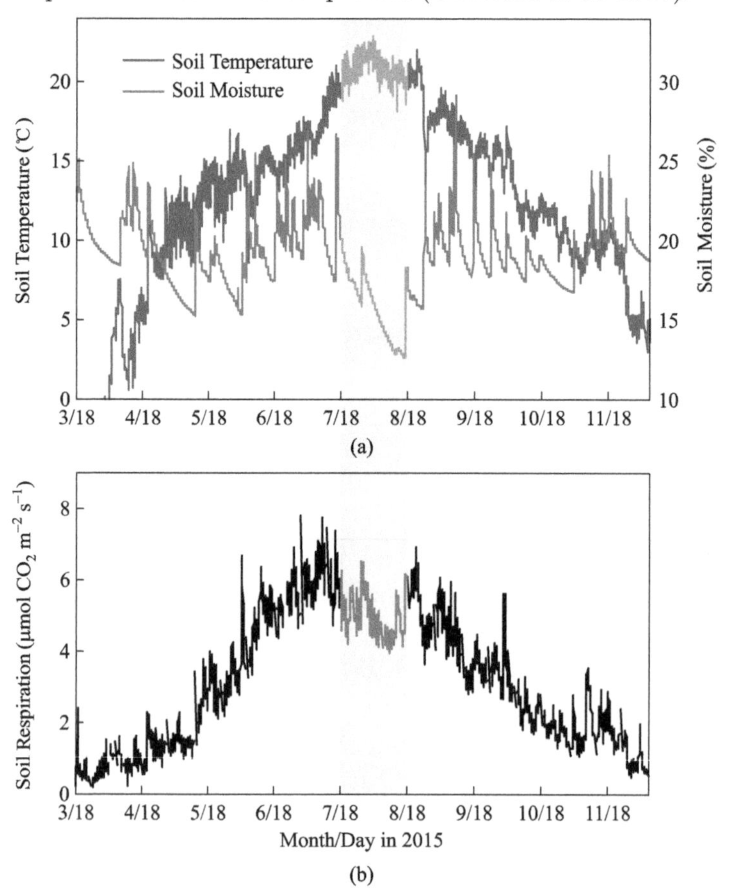

Fig. 3.3 (a) Change in soil temperature (°C) at 5 cm and soil moisture (%) at 10 cm and (b) soil respiration (μmol CO_2 m^{-2} s^{-1}) at a larch forest in the Mt. Fuji Flux Site in central Japan during March 18 and December 17, 2015. The shaded area in yellow indicates a month-long drought period in 2015.

Martin *et al.* (2009) employed a log linear model of multiple variables:

$$\ln(R) = \beta_0 + \beta_1 \cdot T + \beta_2 \cdot T^2 + \beta_3 \cdot \theta + \beta_4 \cdot \theta^2 + \beta_5 \cdot (T \cdot \theta) \tag{3.15}$$

Other variables, such as root biomass or soil C : N ratio can be added to this model. Xu *et al.* (2011), for example, added a parabola function of θ:

$$R = \alpha \cdot e^{\beta_0 \cdot T} \cdot \beta_1 \cdot (\theta - \beta_2)^2 \tag{3.16}$$

where β_1 and β_2 are empirically estimated. Other model forms have also been tested (*see* Table 10.3 in Luo and Zhou 2006). Concilio *et al.* (2005) adopted a slightly different model:

$$R = R_0 \cdot e^{\beta_0 \cdot T} \cdot e^{\beta_1 \cdot \theta} \cdot \beta_2 \cdot T \cdot \theta \tag{3.17}$$

The residuals are further examined as function of litter depth, precipitation, and treatment (*e.g.*, burning, thinning) (*see also* Ma *et al.* 2004, Teramoto *et al.* 2019). At monthly and annual scale, precipitation, or precipitation minus evapotranspiration, can be used to replace soil moisture due to its availability worldwide. Recent studies also found that soil respiration is positively correlated with aboveground photosynthesis. Raich and Schlesinger (1992) proposed a simple linear model for predicting soil respiration from net primary production (NPP), with a linear increasing rate of 1.24.

3.2.7 Time Series Models

Soil and ecosystem respiration change over time due to not only the corresponding changes in significant biophysical variables but also the temporal correlations from memory or legacy effects, especially under extreme climate and disturbances (Besnard *et al.* 2019). It is often necessary to include time in respiration models. This can be done by constructing unique models (*e.g.*, day *vs.* night, by seasons or phenophases), or directly including time as a variable in the model. Considering day of year (DOY) as a critical measure of the seasonal changes in respiration, Xu *et al.* (2011) expanded Equation (3.16) to include an additional quadratic term:

$$R = \alpha \cdot e^{\beta_0 \cdot T} + \beta_1 \cdot (\theta - \beta_2)^2 + \beta_3 \cdot (\mathrm{DOY} - \beta_4)^2 \tag{3.18}$$

This model works well for modeling growing season respiration by assuming a symmetric seasonal pattern before and after the peak respiration (*i.e.*, β_4 value). A more complex form for time series of respiration was proposed by Davidson *et al.* (2006) by using a second-order Fourier regression with DOY:

$$R = \kappa_0 + \kappa_1 \cdot \sin(\text{DOY}^* + \varphi_1) + \kappa_2 \cdot \sin(2 \cdot \text{DOY}^* + \varphi_2) \quad (3.19)$$

where $\text{DOY}^* = \text{DOY} \cdot (2\pi/365)$. Additional higher-order terms can be added to the model (Richardson and Hollinger 2005).

To include all other covariates, with or without consideration of temporal autocorrelations, can be tried with conventional multivariate analysis such as neural network model (Richardson *et al.* 2006). Techniques from Bayesian and machine learning can also be applied to improve the models.

3.3 Measured Datasets for Modeling Soil Respiration

Continuous measurements of soil respiration (μmol CO_2 m^{-2} s^{-1}), soil temperature ($^{\circ}$C) at certain depths such as 5 cm and soil moisture (%) at 10 cm or root zone are used to demonstrate the model applications. The data is from one of the 24 automated chambers (90 cm in length × 90 cm in width × 50 cm in height) installed in a mature larch plantation (*Larix kaempferi*) (35° 26′ 36.7″ N, 138° 45′ 53.0″ E; 1105 m a.s.l.) on the northeastern slope of Mt. Fuji in central Japan. This forest was planted in 1950 and thinned in 2014. Twenty-four automated chambers were installed in 2006 (*see* Fig. 3.1 as an example). Over the course of two hours, the twenty-four chambers were closed sequentially and the sampling period for each chamber was 300 s. CO_2 concentration inside the chamber and the relevant soil and airtemperature and soil moisture were recorded on a datalogger at 10 s intervals. The Chamber #1 is used for demonstrations in this chapter. Understory vegetating inside the chamber had been regularly cleared at about two-week intervals so the measurements represent soil respiration. In 2015, the available data were recorded during the snow-free season between March 18 and December 17. Data gaps of less than 4 hours (*i.e.*, 1 missing value) are filled with the average values before and after the gap, while gaps larger than 4 hours are treated as "no values". Less than 10 extremely high or low respiration measurements (out of 3105) are removed from the database (*see* data and Figs. 3.3 and 3.4 in S3-2: RespirationData.xlsx). During 2013–2017 average (± standard deviation) soil temperature was 8.94 ± 0.23 °C; soil moisture was 17.19 ± 0.97%; and annual total precipitation was 1848 ± 275 mm. Detailed descriptions of the chambers, measurements, data processing, and site characteristics are provided by Teramoto *et al.* (2017, 2019).

As expected in a typical temperate region, soil temperature of the site rapidly increased after March 18, peaked around late July, and decreased linearly to the end of October (Fig. 3.3a). Soil moisture, meanwhile, had episodic changes, *i.e.*, gradually decreasing after each precipitation event. A month-long dry period between July 8 and August 17 had one rainfall, resulting in soil moisture of $< 20\%$. The seasonal changes in soil respi-

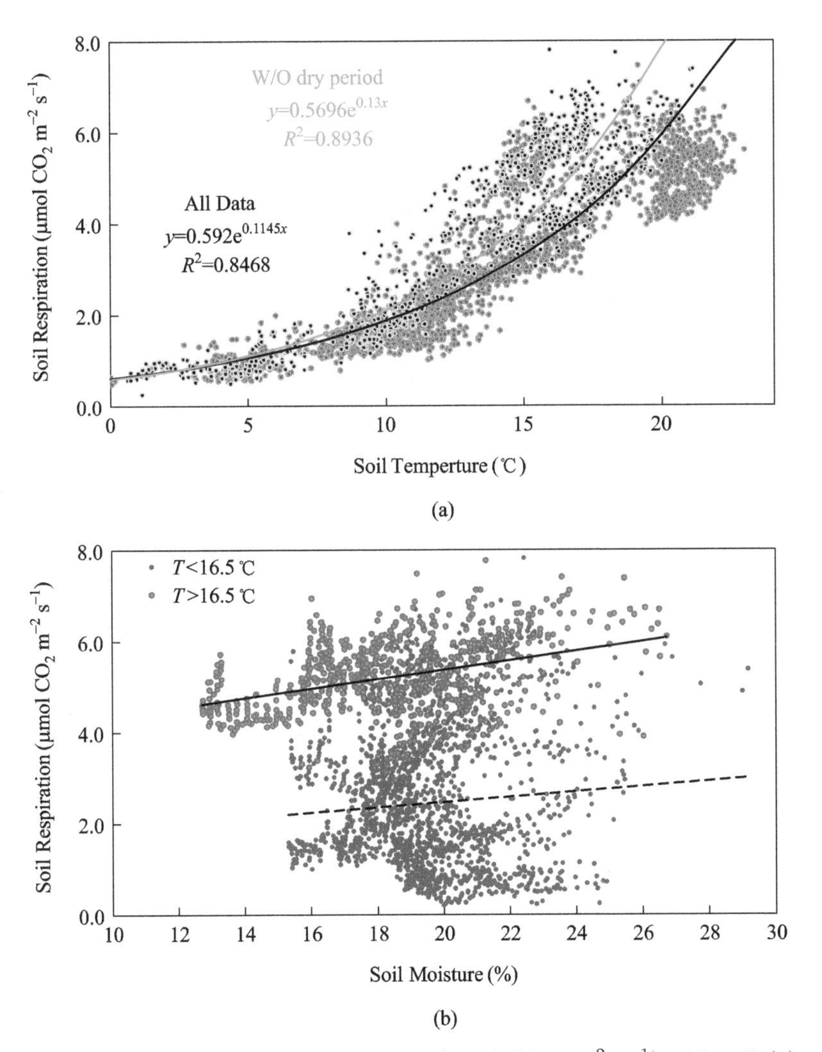

Fig. 3.4 Changes in soil respiration (μmol CO_2 m^{-2} s^{-1}) with soil (a) temperature and (b) moisture. Exponential model (Eq. 3.3) and linear regression lines are presented for the relationships with soil temperature and moisture, respectively. Soil respiration without moisture stress (>20%) is fitted with an independent model (blue).

ration over the measurement period is similar to soil temperature, except during the dry period of July-August. The overall average respiration is 3.19 μmol CO_2 m^{-2} s^{-1}, with values of < 2 μmol CO_2 m^{-2} s^{-1} before May or after October (note that the larch forest generally drops leaves around October 20). Its values are the highest in June (> 6 μmol CO_2 m^{-2} s^{-1}). Respiration during the dry period was reduced by approximately 2 μmol CO_2 m^{-2} s^{-1} (Fig. 3.3a). Another interesting phenomenon is that respiration from late October to mid-November was elevated, which coincided with the relatively high soil temperature and low moisture (Fig. 3.4a).

Soil respiration increased with soil temperature (Fig. 3.4a) and moisture (Fig. 3.4b). Its relationship with soil temperature appeared exponential until soil temperature reached ~16.5 °C at which point it leveled off, or decreased slightly. Exponential model (Eq. 3.3) carried a basal respiration of 0.592 (μmol CO_2 m^{-2} s^{-1}) and an increasing rate of 1.145 μmol CO_2 m^{-2} s^{-1} 10 $°C^{-1}$, yielding a Q_{10} value of 3.14 and an R^2 of 0.847. The variation in respiration also increased with temperature — a typical phenomenon in the literature. For the changes with moisture, there appeared a weak positive relationship (*i.e.*, suppression at lower moisture levels) when soil temperature was greater than 16.5 °C (Fig. 3.5b). The variation in respiration at lower moisture ranges was much higher than that at high moisture ranges. When respiration without water stress was used, the estimated coefficients in the same exponential model had higher slope (1.300 μmol CO_2 m^{-2} s^{-1} 10 $°C^{-1}$), Q_{10} (3.67), and R^2 (0.89) values (Fig. 3.4a). This suggests that it is crucial to include soil moisture in modeling respiration even in this moist forest with an annual precipitation of 1819±252 mm (Teramoto *et al.* 2019).

3.4 Model Performances

To demonstrate model behaviors and performances, models presented in Section 3.2 are applied to the field data (Section 3.3). Their performances are compared in four sets: (1) linear models, (2) nonlinear models, (3) moisture included models, and (4) a DOY-included model.

All models captured the general increasing trends of soil respiration with soil temperature (Fig. 3.5a). The simple linear model showed over-predictions except at low and high respiration and results in a Q_{10} value of 2.7 and an R^2 of 0.752. Predictions from both exponential and quadratic models have obvious over- or under-predictions across the full respiration range, but the variations are high when respiration exceeds 4.0 μmol CO_2 m^{-2}

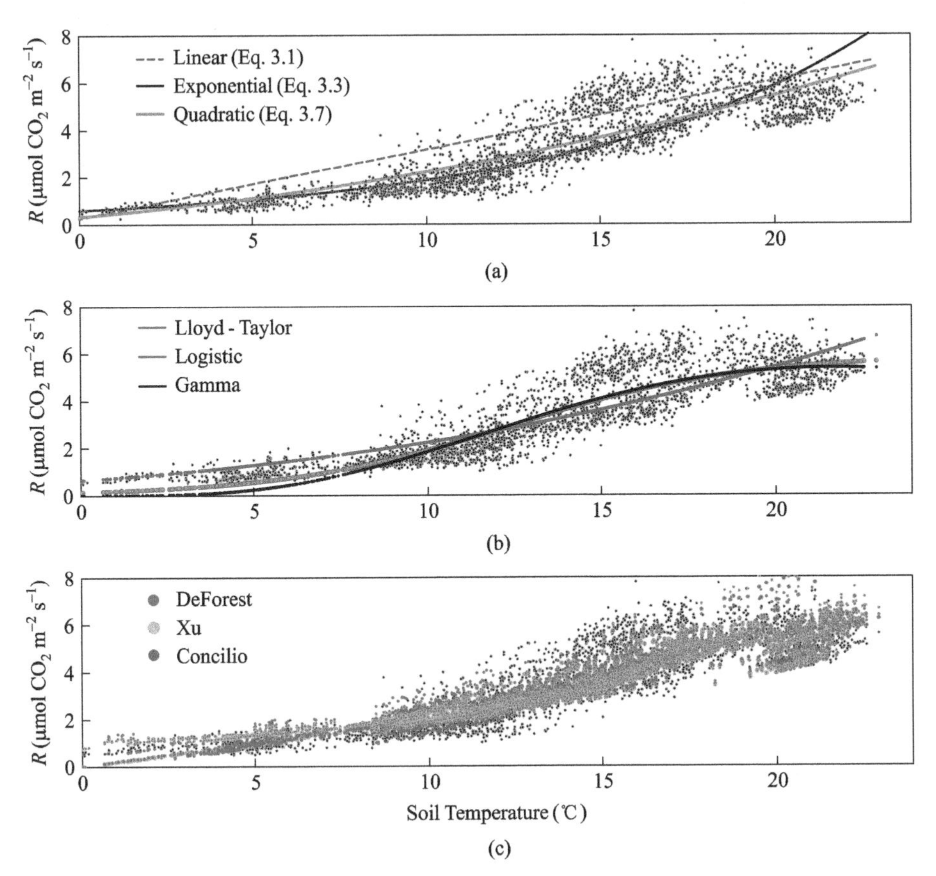

Fig. 3.5 Modeled soil respiration from three sets of models: (a) linear (log-linear) models, (b) nonlinear models, and (c) moisture-included models. Field data were collected at Chamber #1 every 2 hours from March 18 through December 17 in 2015 at a larch plantation in the Mt. Fuji Flux Site, central Japan (Teramoto *et al.* 2019).

s^{-1}. They also have an improved R^2 value at 0.846 and 0.775, respectively (Table 3.1). Logistic and Gamma models have a sigmoid shape, allowing respiration at high temperature to level off or decrease. The Lloyd-Taylor model — the most widely used form in many ecosystem models — performs well and yields a Q_{10} value of 2.9. Moisture-included models have four parameters and predict multiple values at any given temperature. The R^2 values (0.840–0.986) are higher than models solely based on soil temperature. Compared with the mean (standard deviation) value of measured soil respiration (2.746 t C ha^{-1} yr^{-1}), the linear, exponential, and quadratic models produce a mean respiration value of 3.394, 2.683 and 2.753 t C ha^{-1} yr^{-1}, whereas the three nonlinear and moisture-included models produce a

Table 3.1 Statistical estimates of soil respiration with different model forms. The mean and STD are the grand average and standard decimation of the entire sampling ($n = 3105$). α is referred as respiration at 0 °C in some models; and the annual R is modeled annual total carbon loss (t C ha^{-1}) for each model. The measured annual R is 3.192 ($\pm$1.811) t C ha^{-1} yr^{-1}.

Coefficients	Linear	Exponential	Quadratic	Logistic	Lloyd-Taylor	Gamma	DeForest	Xu	Concilio	DOY (Eq. 3.18)
α	0.2489	0.5922	0.3168	5.8164	2.2187	4.5763	4.0896	66.8272	319.9639	99.29090
β_0	0.2886	0.1145	0.1230	3.8513	328.9639	−7.8268	0.0438	0.1004	0.0310	0.04391
β_1			0.0066	−0.3105		−0.2109	0.2178	0.0000	−0.0027	0.00001
β_2							−8.3419	−17.1191	0.00003	−56.15651
β_3										0.00000
β_4										−2721.70967
Mean	3.150	3.120	3.201	3.138	3.220	3.278	3.194	3.230	3.163	3.193
STD	1.687	1.809	1.600	1.740	1.533	1.686	1.659	1.571	1.714	1.674
Annual R	3.394	2.683	2.753	2.699	2.769	2.819	2.747	2.778	2.721	2.746
Q_{10}	2.7	3.1	NA	NA	2.9	NA	NA	NA	NA	NA
R^2	0.752	0.847	0.775	0.82	0.770	0.80	0.840	0.977	0.986	0.854

similar value at ~ 3.2 t C ha^{-1} yr^{-1}. The overestimation of 0.649 t C ha^{-1} yr^{-1} from the linear model accounts 23.6% of the actual measurement (Fig. 3.6).

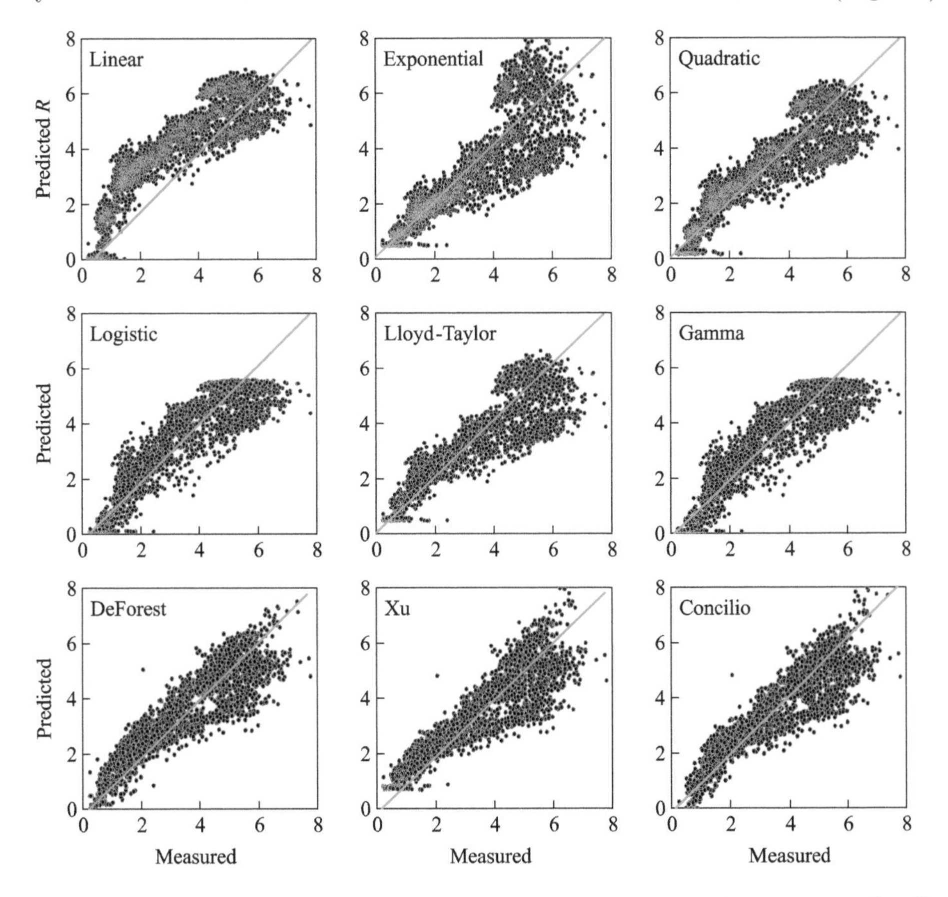

Fig. 3.6 Comparisons of predicted and measured soil respiration (μmol CO_2 m^{-2} s^{-1}) from 9 models (Fig. 3.5). The cyan lines are the 1 : 1 ratios. Field data were collected with an automated respiration chamber every 2 hours from March 18 through December 17 in 2015 at a larch plantation in the Mt. Fuji Flux Site, Japan (Teramoto *et al.* 2019).

Including day of year (DOY) in a respiration model (Eq. 3.18) requires six parameters to be estimated through nonlinear regression analysis (S3-2). It does not improve the model's overall performance (R^2=0.854), but it produces the closest annual total estimate (3.193 t C ha^{-1} yr^{-1}) to the actual measurement and captures many unique changes throughout the year(Fig. 3.7), such as the decreases during the dry period of July 18–August 17 when compared to predicted respiration from Lloyd-Taylor model. However, it does not capture the high respiration in June and the low respiration in late April. Finally, data from one of the 24 chambers are used in this chapter. It is well known that both soil microclimate and vegetation chara-

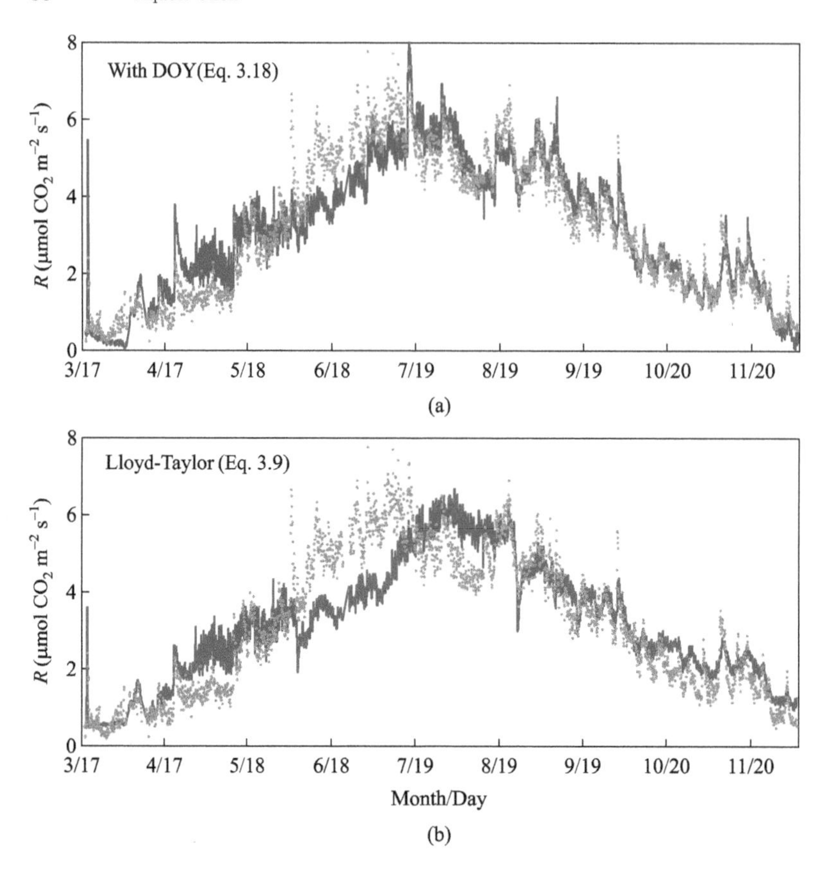

Fig. 3.7 Comparisons of predicted and measured soil respiration (μmol CO_2 m^{-2} s^{-1}) between (a) DOY-included model and (b) Lloyd-Taylor model. Field measurements (gray dots) were collected with an automated respiration chamber every 2 hours from March 18 through December 17 in 2015 at a larch plantation in the Mt. Fuji Flux Site, central Japan (Teramoto *et al.* 2019).

cteristics within a forest community vary substantially (Chen *et al.* 1999). To accurately predict the ecosystem-level respiration, independent models are needed for each chamber. Using stand average values of independent variables for predictions is not recommended.

3.5 Summary

Accurate prediction of ecosystem respiration remains a challenge because of lack of theoretical foundation. Modeling ecosystem respiration is mostly based on Van't Hoff's theory where temperature is the sole driving force for changes in respiration. Q_{10} value, which describes increase in rate of reaction when temperature is raised by 10 °C, is widely used for model development but has been challenged by studies in the literature, where several dozen forms have been exercised. Among these models, the temperature-dependent Q_{10} model is mostly widely applied in ecosystem models because of its simplicity (Chapter 2). Including available soil water in the model is nonetheless critical in modeling soil respiration. For modeling the changes of respiration, inclusions of phenophase and DOY of year appear necessary. More importantly, applying the same model for diverse ecosystem types or under very different climatic conditions should be cautiously practiced because no single model will work for all kinds of conditions. This is because these models are all empirical, with their coefficients estimated with *in-situ* measurements of respiration and forcing variables. Field measurements of soil respiration, soil temperature and moisture have been used for these model developments and validations. Based on the literature and model performances, the following aspects are critical for development of reliable respiration models:

- Simple linear, power, and polynomial forms are not recommended in modeling respiration regardless of their simple-to-use nature.
- Selection of model form is critical for producing reliable predictions. Residual analysis can help development of additional covariates and model forms.
- Incorporating other independent variables is necessary. Soil moisture, soil carbon and nutrient content, biomass or production, canopy cover, litter depth, *etc.* are among the potential factors to be considered.
- Multiple model forms or a unique set of coefficients for the same model need to be used for different times such as seasons (phenophases), climatic conditions, and disturbances. For modeling seasonal changes, day of year should be included in the models.

Online Supplementary Materials

S3-1: Spreadsheet models (Schematics.xlsx) for illustrating the roles of two parameters in exponential model (Eq. 3.3) for respiration-temperature relationship, calculations of Q_{10} values, and inclusion of linear constraints by moisture (θ) at high temperature ranges (Fig. 3.3).

S3-2: Field measurements of soil respiration, soil temperature and moisture in 2015 from Chamber #1 (RespirationData.xlsx) in a mature larch plantation (*Larix kaempferi*) (35° 26′ 36.7″ N, 138° 45′ 53.0″ E; 1105 m a.s.l.) on the northeastern slope of Mt. Fuji in central Japan (Fig. 3.3).
S3-3: Spreadsheet modeling and model comparisons (Rmodel_1.xlsx) of linear, exponential and quadratic forms (Eqs. 3.5, 3.3, 3.7).
S3-4: Spreadsheet modeling and model comparisons (Rmodel_2.xlsx) of Logistic, Lloyd-Taylor and Gamma models (Eqs. 3.10, 3.9, 3.11).
S3-5: Spread sheet modeling and model comparisons (Rmodel_3.xlsx) of three model forms by including soil moisture (θ) as an additional independent variable (Eqs. 3.14, 3.16, 3.17).
S3-6: Spreadsheet modeling of soil respiration with day of year (DOY) and soil moisture (θ) as additional covariates of temperature (Rmodel_4.xlsx) (Eq. 3.18) (Fig. 3.7).
S3-7: Python codes for estimating empirical coefficients through nonlinear regression analysis of Logistic, Lloyd-Taylor, Gamma, DeForest, Xu, Concilio and DOY models (Respiration.rar). This file has two Excel data files and 12 Python programs for linear and non-linear regression.

Scan the QR code or go to
https://msupress.org/supplement/BiophysicalModels
to access the supplementary materials.
User name: biophysical
Password: z4Y@sG3T

Acknowledgements

The author appreciates Dr. Naishen Liang of National Institute for Environmental Studies, Japan for sharing his field measurements and hosting a field visit to the forest in Mt. Fuji. Fruitful discussion on respiration studies was made with many LEES members and collaborators, including Amy Concilio, Jared DeForest, Eugenie Euskirchen, Siyan Ma, James Le Moine, Asko Noormets, Malcolm North, Qinglin Li, Xianglan Li, Jianye Xu, and others. Huimin Zou developed Python codes for the non-linear regression analysis. Marium Altaf Arain, Jared DeForest, and Naishen Liang provided insightful reviews on early drafts of the chapter. Kristine Blakeslee edited the language and format of the draft.

References

Barr, A. G., Griffis, T. J., Black, T. A., Lee, X., Staebler, R. M., Fuentes, J. D., Chen, Z., and Morgenstern, K. (2002). Comparing the carbon budgets of boreal and temperate deciduous forest stands. *Canadian Journal of Forest Research*, 32(5), 813–822.

Besnard, S., Carvalhais, N., Arain, M. A., Black, A., Brede, B., Buchmann, N., Chen, J., Clevers, J. G. P. W., Dutrieux, L. P., Gans, F., Herold, J. M., Kosugi, Y., Knohl, L., Law, B. E., Paul-Limoges, E., Lohila, A., Merbold, L., Roupsard, O., Valentini, R., Wolf, S., Zhang, X., and Reichstein, H. M. (2019). Memory effects of climate and vegetation affecting net ecosystem CO_2 fluxes in global forests. *PloS One*, 14(2), e0211510.

Chen, J., John, R., Sun, G., McNulty, S., Noormets, A., Xiao, J., Turner, M. G., and Franklin, J. F. (2014). Carbon fluxes and storage in forests and landscapes. In *Forest Landscapes and Global Change* (pp. 139–166). Springer.

Chen, J., Saunders, S. C., Crow, T. R., Naiman, R. J., Brosofske, K. D., Mroz, G. D., Brookshire, B. L., and Franklin, J. F. (1999). Microclimate in forest ecosystem and landscape ecology: Variations in local climate can be used to monitor and compare the effects of different management regimes. *BioScience*, 49(4), 288–297.

Concilio, A., Ma, S., Li, Q., LeMoine, J., Chen, J., North, M., Moorhead, D., and Jensen, R. (2005). Soil respiration response to prescribed burning and thinning in mixed-conifer and hardwood forests. *Canadian Journal of Forest Research*, 35(7), 1581–1591.

Contosta, A. R., Burakowski, E. A., Varner, R. K., and Frey, S. D. (2016). Winter soil respiration in a humid temperate forest: The roles of moisture, temperature, and snowpack. *Journal of Geophysical Research: Biogeosciences*, 121(12), 3072–3088.

Curiel Yuste, J., Baldocchi, D. D., Gershenson, A., Goldstein, A., Misson, L., and Wong, S. (2007). Microbial soil respiration and its dependency on carbon inputs, soil temperature and moisture. *Global Change Biology*, 13(9), 2018–2035.

Davidson, E. A., Belk, E., and Boone, R. D. (1998). Soil water content and temperature as independent or confounded factors controlling soil respiration in a temperate mixed hardwood forest. *Global Change Biology*, 4(2), 217–227.

Davidson, E. A., Janssens, I. A., and Luo, Y. (2006). On the variability of respiration in terrestrial ecosystems: Moving beyond Q_{10}. *Global Change Biology*, 12(2), 154–164.

DeForest, J. L., Noormets, A., McNulty, S. G., Sun, G., Tenney, G., and Chen, J. (2006). Phenophases alter the soil respiration-temperature relationship in an oak-dominated forest. *International Journal of Biometeorology*, 51(2), 135–144.

Euskirchen, E. S., Chen, J., Gustafson, E. J., and Ma, S. (2003). Soil respiration at dominant patch types within a managed northern Wisconsin landscape. *Ecosystems*, 6(6), 595–607.

Falge, E., Tenhunen, J., Baldocchi, D., Aubinet, M., Bakwin, P., Berbigier, P., Bernhofer, C., Bonnefond, J-C., Burba, G., Clement, R., Davis, K. J., Elbers, J. A., Falk, M., Goldstein, A. H., Grelle, A., Granier, A., Grünwald, T., Guðmundsson, J., Hollinger, D., Janssens, I. A., Keronen, P., Kowalski, A. S., Katul, G., Law, B. E., Malhi, Y., Meyers, T., Monson, R. K., Moors, E., Munger, J. W., Oechel, W., Paw U, K. T., Pilegaard, K., Rannik, Ü., Rebmann, C., Suyker, A., Thorgeirsson,

H., Tirone, G., Turnipseed, A., Wilson, K., and Wofsy, S. (2002). Phase and amplitude of ecosystem carbon release and uptake potentials as derived from FLUXNET measurements. *Agricultural and Forest Meteorology*, 113(1–4), 75–95.

Giasson, M. A., Ellison, A. M., Bowden, R. D., Crill, P. M., Davidson, E. A., Drake, J. E., and Munger, J. W. (2013). Soil respiration in a northeastern US temperate forest: A 22-year synthesis. *Ecosphere*, 4(11), 1–28.

Görres, C. M., Kammann, C., and Ceulemans, R. (2016). Automation of soil flux chamber measurements: Potentials and pitfalls. *Biogeosciences*, 13(6), 1949–1966.

Gough, C. M., Vogel, C. S., Kazanski, C., Nagel, L., Flower, C. E., and Curtis, P. S. (2007). Coarse woody debris and the carbon balance of a north temperate forest. *Forest Ecology and Management*, 244(1–3), 60–67.

Harmon, M. E., Bond-Lamberty, B., Tang, J., and Vargas, R. (2011). Heterotrophic respiration in disturbed forests: A review with examples from North America. *Journal of Geophysical Research: Biogeosciences*, 116(G4). http://doi.org/10.1029/2010JG001495.

Hill, N. B., Riha, S. J., and Walter, M. T. (2018). Temperature dependence of daily respiration and reaeration rates during baseflow conditions in a northeastern US stream. *Journal of Hydrology-Regional Studies*, 19, 250–264.

Högberg, P., Nordgren, A., Buchmann, N., Taylor, A. F., Ekblad, A., Högberg, M. N., Nyberg, G., Ottosson-Löfvenius, M., and Read, D. J. (2001). Large-scale forest girdling shows that current photosynthesis drives soil respiration. *Nature*, 411(6839), 789–792.

Khomik, M., Arain, M. A., Liaw, K. L., and McCaughey, J. H. (2009). Debut of a flexible model for simulating soil respiration–soil temperature relationship: Gamma model. *Journal of Geophysical Research: Biogeosciences*, 114(G3), doi: 10.1029/2008JG000851.

Khomik, M., Arain, M. A., and McCaughey, J. H. (2006). Temporal and spatial variability of soil respiration in a boreal mixedwood forest. *Agricultural and Forest Meteorology*, 140(1–4), 244–256.

Lavigne, M. B., Boutin, R., Foster, R. J., Goodine, G., Bernier, P. Y., and Robitaille, G. (2003). Soil respiration responses to temperature are controlled more by roots than by decomposition in balsam fir ecosystems. *Canadian Journal of Forest Research*, 33(9), 1744–1753.

Lavigne, M. B., Franklin, S. E., and Hunt Jr, E. R. (1996). Estimating stem maintenance respiration rates of dissimilar balsam fir stands. *Tree Physiology*, 16(8), 687–695.

Law, B. E., Ryan, M. G., and Anthoni, P. M. (1999). Seasonal and annual respiration of a ponderosa pine ecosystem. *Global Change Biology*, 5(2), 169–182.

Leather, J. W. (1915). Soil gases. *Memoirs of the Department of Agriculture,* 4(3), 85–134.

Li, Q., Chen, J., and Moorhead, D. L. (2012). Respiratory carbon losses in a managed oak forest ecosystem. *Forest Ecology and Management*, 279, 1–10.

Lieth, H., and Ouellette, R. (1962). Studies on the vegetation of the Qaspé peninsula: ii. The soil respiration of some plant communities. *Canadian Journal of Botany*, 40(1), 127–140.

Lloyd, J., and Taylor, J. A. (1994). On the temperature dependence of soil respiration. *Functional Ecology*, 8(3),315–323.

Lundegårdh, H. (1922). Neue apparate zur analyse des kohlensaure der luft. *Biochem. Ztschr.* 131, 109.

Lundegårdh, H. (1927). Carbon dioxide evolution of soil and crop growth. *Soil Science*, 23(6), 417–453.

Luo. Y., and Zhou, X. (2006). *Soil Respiration and the Environment.* Elsevier. 328pp.

Ma, S., Chen, J., North, M., Erickson, H. E., Bresee, M., and Le Moine, J. (2004). Short-term effects of experimental burning and thinning on soil respiration in an old-growth, mixed-conifer forest. *Environmental Management*, 33(1), S148–S159.

Martin, J. G., Bolstad, P. V., Ryu, S. R., and Chen, J. Q.(2009). Modeling soil respiration based on carbon, nitrogen, and root mass across diverse Great Lake forests. *Agricultural and Forest Meteorology*, 149(10), 1722–1729.

Mahecha, M. D., Reichstein, M., Carvalhais, N., Lasslop, G., Lange, H., Seneviratne, S. I., Vargas, R., Ammann, C., Arain, M. A., Cescatti, A., Janssens, I. A., Migliavacca, M., Montagnani, L., and Richardson, A. D. and Janssens, I. A. (2010). Global convergence in the temperature sensitivity of respiration at ecosystem level. *Science*, 329(5993), 838–840.

Meyer, N., Welp, G., and Amelung, W. (2018). The temperature sensitivity (Q_{10}) of soil respiration: Controlling factors and spatial prediction at regional scale based on environmental soil classes. *Global Biogeochemical Cycles*, 32(2), 306–323.

Neller, J. R. (1918). Studies on the correlation between the production of carbon dioxide and the accumulation of ammonia by soil organisms. *Soil Science*, 5, 225–242.

Perkins, D. M., Yvon-Durocher, G., Demars, B. O., Reiss, J., Pichler, D. E., Friberg, N., Trimmer, M., and Woodward, G. (2012). Consistent temperature dependence of respiration across ecosystems contrasting in thermal history. *Global Change Biology*, 18(4), 1300–1311.

Pettenkofer, M. (1861). *Ueber einen neuen Respirations-Apparat.* K. Akademie, in Commiss. bei G. Franz.

Petersen, P. (1870). *Ueber den Einfluss des Mergels auf die Bildung von Kohlensäure und Salpetersäure im Ackerboden. Landwirtschaftliche* Versuchsstationen. Bd. *XIII.* 1870. S. 155-175.

Raich, J. W., and Nadelhoffer, K. J. (1989). Belowground carbon allocation in forest ecosystems: Global trends. *Ecology*, 70(5), 1346–1354.

Raich, J. W., and Schlesinger, W. H. (1992). The global carbon dioxide flux in soil respiration and its relationship to vegetation and climate. *Tellus B*, 44(2), 81–99.

Reichstein, M., Falge, E., Baldocchi, D., Papale, D., Aubinet, M., Berbigier, P., Bernhofer, C., Buchmann, N., Gilmanov, T., Granier, A., Grünwald, T., Havrávanková, K., Ilvesniemi, H., Janous, D., Knohl, A., Laurila, T., Lohila, A., Loustau, D., Matteucci, G., Meyers, T., Miglietta, F., Ourcival, J-M., Pumpanen, J., Rambal, S., Rotenberg, E., Sanz, M., Tenhunen, J., Seufert, G., Vaccari, F., Vesala, T., Yakir, D., and Valentini, R. (2005). On the separation of net ecosystem exchange into assimilation and ecosystem respiration: Review and improved algorithm. *Global Change Biology*, 11(9), 1424–1439.

Richardson, A. D., Braswell, B. H., Hollinger, D. Y., Burman, P., Davidson, E. A., Evans, R. S., Flanagan, L. B., Munger, J. W., Savage K., Urbanski, S. P., and Wofsy, S. C. (2006). Comparing simple respiration models for eddy flux and dynamic chamber data. *Agricultural and Forest Meteorology*, 141(2–4), 219–234.

Richardson, A. D., and Hollinger, D. Y. (2005). Statistical modeling of ecosystem respiration using eddy covariance data: Maximum likelihood parameter estimation, and Monte Carlo simulation of model and parameter uncertainty, applied to three simple models. *Agricultural and Forest Meteorology*, 131(3–4), 191–208.

Ryan, M. G., and Law, B. E. (2005). Interpreting, measuring, and modeling soil respiration. *Biogeochemistry*, 73(1), 3–27.

Schlesinger, W. H. (1977). Carbon balance in terrestrial detritus. *Annual Review of Ecology and Systematics*, 8(1): 51–81.

Teramoto, M., Liang, N., Zeng, J., Saigusa, N., and Takahashi, Y. (2017). Long-term chamber measurements reveal strong impacts of soil temperature on seasonal and inter-annual variation in understory CO_2 fluxes in a Japanese larch (*Larix kaempferi* Sarg.) forest. *Agricultural and Forest Meteorology*, 247, 194–206.

Teramoto, M., Liang, N., Takahashi, Y., Zeng, J., Saigusa, N., Ide, R., and Zhao, X. (2019). Enhanced understory carbon flux components and robustness of net CO_2 exchange after thinning in a larch forest in central Japan. *Agricultural and Forest Meteorology*, 274, 106–117.

Van't Hoff, J. H. (1884). *Etudes de Dynamique Chimique*. Frederik Muller & Company.

Van't Hoff, J. H. (1898). *Vorlesungen über theoretische und physikalische Chemie*. Vieweg.

Van Gorsel, E., Delpierre, N., Leuning, R., Black, A., Munger, J. W., Wofsy, S., Aubinet, M., Feigenwinter, C., Beringer, J., Bonal, D., Chen, B., Chen, J., Clement, R., Davis, M., Desai, A. R., Dragoni, D., Etzold, S., Grünwald, T., Gu, L., Heinesch, B., Hutyra, L. R, Jans, W. W. P., Kutsch, W., Law, B. E., Leclerc, M. Y., Mammarella, I, Montagnani, L., Noormets, A., Rebmann, C., and Wharton, S. and Chen, B. (2009). Estimating nocturnal ecosystem respiration from the vertical turbulent flux and change in storage of CO_2. *Agricultural and Forest Meteorology*, 149(11), 1919–1930.

Wang, C., Han, Y., Chen, J., Wang, X., Zhang, Q., and Bond-Lamberty, B. (2013). Seasonality of soil CO_2 efflux in a temperate forest: Biophysical effects of snowpack and spring freeze-thaw cycles. *Agricultural and Forest Meteorology*, 177, 83–92.

Wollny, E. (1818). *Forschungen auf dem Gebiete der Agrikultur-Physik*. Carl Winter's Universitätsbuchhandlung Heidelberg. Bd. 1–20, 1878–1898.

Wollny, E. (1880). Untersuchungen über den Kohlensäuregehalt der Bodenluft. *Landwirtschaftliche Versuchs-Stationen*, 25, 373–391.

Wofsy, S. C., Goulden, M. L., Munger, J. W., Fan, S. M., Bakwin, P. S., Daube, B. C., Bassow, S. L., and Bazzaz, F. A. (1993). Net exchange of CO_2 in a mid-latitude forest. *Science*, 260(5112), 1314–1317.

Xu, J., Chen, J., Brosofske, K., Li, Q., Weintraub, M., Henderson, R., Wilske, B., John, R., Jensen, R., Li, H., and Shao, C. (2011). Influence of timber harvesting alternatives on forest soil respiration and its biophysical regulatory factors over a 5-year period in the Missouri Ozarks. *Ecosystems*, 14(8), 1310–1327.

Xu, L. K., Baldocchi, D. D., and Tang, J. (2004). How soil moisture, rain pulses, and growth alter the response of ecosystem respiration to temperature. *Global Biogeochemical Cycles*, 18(4), doi:10.1029/2004GB002281.

Xu, M., and Qi, Y. (2001). Soil-surface CO_2 efflux and its spatial and temporal variations in a young ponderosa pine plantation in northern California. *Global Change Biology*, 7(6), 667–677.

Yakir, D., and da SL Sternberg, L. (2000). The use of stable isotopes to study ecosystem gas exchange. *Oecologia*, 123(3), 297–311.

Yu, X., Zha, T., Pang, Z., Wu, B., Wang, X., Chen, G., Li, C., Cao, J., Jia, G., Li, X., and Wu, H. (2011). Response of soil respiration to soil temperature and moisture in a 50-year-old oriental arborvitae plantation in China. *PloS One*, 6(12), e28397.

Yuan, W. P., Luo, Y. Q., Li, X. L., Liu, S. G., Yu, G. R., Zhou, T., Bahn, M., Black, A., Desai, A. R, Cescatt i, A, Marcolla, B., Jacobs, C., Chen, J, Aurela, M., Bemhofer, C., Gielen, B, Bohrer, G, Cook, D. R, Dragoni, D., Dunn, A. L., Gianelle, D, Grünwald, T., Ibrom, A., Leclerc, M. Y., Lindroth, A. Liu, H., Marchesini, L.B., Montagnani, L., Pita, G., Rodeghiero, M., Rodrigues, A., Star, G., and Stoy, P. C. and Marcolla, B. (2011). Redefinition and global estimation of basal ecosystem respiration rate. *Global Biogeochemical Cycles*, 25(4), doi:10.1029/2011GB004150.

Chapter 4
Modeling Evapotranspiration

Ge Sun and Jiquan Chen

4.1 Introduction

The process by which water changes from liquid form to gas from a surface is called evaporation — a term used since the mid-1600s according to Stanhill (2005). Water can also move directly from the snowpack to the atmosphere due to the direct phase transition of snow to water vapor, which is called snow sublimation and is considered part of evaporation in the literature (Stigter *et al.* 2018). In terrestrial ecosystems where vegetation dominates land surfaces, the processes of transporting water into the atmosphere from bare lands (*i.e.*, soil, rock and paved surfaces), water (*i.e.*, streams, ponds and open-water swamps), and vegetation-covered areas is called evapotranspiration (ET) — a term created by American geographer/climatologist Charles Warren Thornthwaite (1948). While devising a climate classification system that is still used worldwide (Hewlett 1982), Thornthwaite proposed "potential evapotranspiration" as an "index of thermal efficiency" in predicting crop growth. Here the water loss through plant growth

Ge Sun
Eastern Forest Environmental Threat Assessment Center, USDA Forest Service
Email: ge.sun@usda.gov

Jiquan Chen
Landscape Ecology & Ecosystem Science (LEES) Lab, Department of Geography, Environment, and Spatial Sciences & Center for Global Change and Earth Observations, Michigan State University, East Lansing, MI 48823
Email: jqchen@msu.edu

Jiquan Chen, *Biophysical Models and Applications in Ecosystem Analysis*,
https://doi.org/10.3868/978-7-04-055256-0-4

is called transpiration to distinguish it from water loss through evaporation. He argued that "We cannot tell whether a climate is moist or dry by knowing the precipitation alone. We must know whether precipitation is greater or less than the water needed for evaporation and transpiration." Potential evapotranspiration (PET, cm), a term used since 1937 (Stanhill 2005), is calculated with air temperature (°C) by Thornthwaite, with a unit that is the same as precipitation (*e.g.*, cm):

$$\mathrm{PET} = c \cdot T^{a} \tag{4.1}$$

where T (°C) is the monthly mean air temperature, and c and a are empirical parameters that are hypothesized as a function of heat index (Thornthwaite 1948). Although modern terms (*e.g.*, vapor pressure deficit) were not used in Thornthwaite's paper, heat stress and precipitation were already considered in modeling PET.

In some regions and disciplines, the term *evaporation* is preferred over *evapotranspiration* (Savenije 2004). Penman (1963) objected to the use of the term *evapotranspiration* because evaporation already includes transpiration, since the latter is an evaporation process from leaf surface (Stanhill 2005). In a mathematical form, ecosystem ET includes three sub-components:

$$\mathrm{ET} = T + I + E \tag{4.2}$$

where T is vegetation transpiration, I is evaporation from canopy interception, and E is evaporation from soil and vegetation surfaces (*e.g.*, stems, standing and downed logs). All terms are normally expressed in millimeters (mm) but can be quantified in mass weight (*e.g.*, kg) or energy by multiplying ground surface area. One mm of ET equals to 1.0 g of water (*i.e.*, water density 1 gram per milliliter). The amount of energy needed to evaporate water is called latent heat of vaporization of water (λ, J $\mathrm{mol^{-1}}$), which varies by temperature and atmospheric pressure. At air pressure of 101 kPa, its value is 44.6 kJ $\mathrm{mol^{-1}}$at 10 °C and 44.1 kJ $\mathrm{mol^{-1}}$at 20 °C (note the molecular mass of H_2O is 18 g $\mathrm{mol^{-1}}$, and 1 watt (W) = 1.00 joules per second (J $\mathrm{s^{-1}}$)). ET from mass balance (Eq. 4.2) is approached as precipitation minus stream flow, change in soil moisture and exchange with the deep ground water. From an energy balance perspective, ET (W $\mathrm{m^{-2}}$) can be calculated as:

$$\mathrm{ET} = R_{\mathrm{n}} - H - G - \Delta S \tag{4.3}$$

where R_{n} is net radiation, H is the sensible heat, G is the soil heat flux, and ΔS is the heat storage over a period of time within the canopy column (air and vegetation). All terms have a unit of W $\mathrm{m^{-2}}$ (*see* Chapter 1). The sum of ($R_n - G - \Delta S$) is equal to the sum of H and ET, which is termed as available energy in the literature. When the unit of ET is millimeters, the left side of the equation is replaced with λET. Clearly, the mass balance in Equation (4.2) emphasizes the subcomponents of ET, whereas the Equation (4.3) model shows ET from an energy balance perspective. The mass balance approach has been favored in the hydrological community, while the energy balance approach appears to be preferred in micrometeorology and ecosystem studies, though both methods are employed in all scientific disciplines.

Recent global synthesis on ET partitions into transpiration and other components suggests plant transpiration dominates the ET flux for most vegetated surfaces (Jasechko *et al.* 2013), while ($I+E$) can be as high as 30% of total ET. ($I+E$) can be an important component of ecosystem ET in very dry or very wet environments. These three components (T, I, and E) must be estimated independently for effective water management and water-ecosystem interactions.

Ecologists and hydrologists are interested in understanding and quantifying ET from several perspectives:

1. Ecosystem processes. ET is a key variable that is directly coupled with ecosystem productivity and carbon sequestration (Aber and Federer 1992). This is easy to understand because carbon dioxide (CO_2) uptake during plant photosynthesis uses the same pores (stomata) as water loss (transpiration) pathways (Gedney *et al.* 2006). Evapotranspiration is the only variable that links hydrology and biological processes in many ecosystem models. ET is also highly linked to ecosystem productivity and net ecosystem exchange of CO_2 between an ecosystem and the atmosphere because photosynthesis and ecosystem respiration are controlled by available energy and soil water availability (*see* Chapters 1 and 2).
2. Water balances. ET is a large component of the water budget at global scale. Worldwide, mean annual ET rates are estimated to be about 600 mm or 60%–70% of precipitation (Oki and Kanae 2006). In the United States more than 70% of annual precipitation returns to the atmosphere as ET (Sanford and Selnick 2013), while the percentage is high as 90% in Australia (McMahon *et al.* 2013). Regional annual ET rates can be as high as 85% of precipitation in forest landscapes in the humid southern United States (Lu *et al.* 2003). Vegetation affects watershed water balances by influencing ET (Zhang *et al.* 2001, Bosch and Hewlett 1982), though its influence is minor compared to climate (Oudin *et al.* 2008).

3. Emerging global change science. ET is a key variable in meteorological, hydrological, and ecosystem sciences (Baldocchi *et al.* 2000); it must be considered in contemporary watershed management (Sun *et al.* 2008a). Climate change and land use change directly affect the hydrological cycle and water resources through altering ET processes (Sun *et al.* 2011a). For example, an increase in air temperature generally means an increase in evaporative demands, resulting in an increase in water loss through ET — thus a decrease in groundwater recharge, available soil water, stream flow, and human water supply. Similarly, land use conversion (*i.e.*, bioenergy crop expansion, deforestation) can dramatically change plant cover and biomass, affecting transpiration and evaporation rates, site water balance, and regional water resources. Climate change and land use change directly affect the hydrological cycle and water resources through altering the ET processes (Caldwell *et al.* 2012). Changes in water balance patterns will consequently impact stream flow and water quality. Accurate quantifications of ET would better prepare us respond to climate change and its many consequences (Vose *et al.* 2011). Recent decades have witnessed rapid growth in remote sensing technology and other "big data" for cross-disciplinary "big science" (*e.g.*, Schellekens *et al.* 2017). This growth is fueled by modern computing capability, permitting us to re-evaluate the spatiotemporal changes in ET and associated components.
4. Climate change and feedbacks. ET is the key link between energy, water balances, and climate systems. More than half of the solar radiation absorbed by the land surface (Trenberth *et al.* 2009) is used for ET. Changes in ET directly affect runoff, soil water, and local precipitation (Koster *et al.* 2004), temperature, and local-to-region air humidity (Hao *et al.* 2018). ET is tightly coupled to land surface energy balance (Chen *et al.* 2004) in both managed and unmanaged systems (Sun *et al.* 2010), and thus influences vegetation-climate feedbacks (Bonan 2008). For example, studies have found that reforestation in China lowered local air temperature due to increased ET through reforestation (Peng *et al.* 2014).
5. Ecosystem diversity. ET is also used as an index to represent the available environmental energies and ecosystem productivity. For example, Currie (1991) found that 80%–93% of the variability in species richness in the four vertebrate classes that the author studied could be statistically explained by a monotonically increasing function of a single variable: potential evapotranspiration. In contrast, tree richness was more closely related to actual ET.

4.2 Methods for Quantifying ET

ET may be directly measured or estimated using models discussed later. Field methods for measurement are diverse and have been evolving (Table 4.1). For a specific ecosystem, Stanhill (2005) categorized these methods into three major groups: lysimeter measurement, Bowen ratio method, and eddy-covariance method. Prior to modern eddy-covariance flux towers, the Lagarangian profiling method was also used by meteorologists (Chapter 1) Campbell and Norman 2012). At landscape and watershed scale, catchment water balance is examined in hydrological studies because ET is the residual of (precipitation–stream flow–change in soil water) by ignoring the water movement between land surface and ground water that is small (especially at hourly to days scale; Hewlett 1982). During a short period of time (*e.g.*, hours to days) when there is no precipitation, the changes in soil water can be used to approximate ET by the water balance principle. The past two decades also witnessed increasing use of remote sensing spectra to indirectly measure (*i.e.*, remote sensing modeling) ET at regional to global scales (*e.g.*, Mu *et al.* 2007, Jung *et al.* 2010).

ET is complex in nature, involving both physical and physiological processes that vary tremendously in space and time (Shuttleworth 2012, Amatya *et al.* 2014). Accurate quantification of ET for long periods and large areas is often costly, if possible at all; thus ET remains one of the least-measured components of the hydrologic cycle. ET is still an imprecise science (Shuttleworth 2012). There are many ways to quantify ET at different scales (Table 4.1). Direct ecosystem-scale ET measurement techniques include catchment water balance (Bosch and Hewlett 1982), sap flow (Smith and Allen 1996), eddy-covariance (Baldocchi *et al.* 1988) and Bowen ratio (Bowen 1926) methods. It is worth noting that the Bowen ratio (*i.e.*, sensible heat to latent heat ratio) was not used to measure ET until the 1950s (Webb 1960). Remote sensing techniques allow monitoring ET at a very large scale, but the estimated values are snapshots of time series at low frequencies, and modeling is involved to extrapolate for continuous estimation. Wilson *et al.* (2001) and Domec *et al.* (2012) compared multiple ET methods and found that each had its own advantages and limitations. For example, the watershed water balance method, which is typically applicable to long-term average ET estimates, has errors when change in soil moisture is ignored. ET would be overestimated or underestimated when ET is computed as residual of precipitation and runoff. Sap flow measurements provide a powerful tool for quantifying plant water use and physiological responses of plants

Table 4.1 A comparison of major methods for estimating evapotranspiration (ET).

	Methods	Strengths	Weaknesses	Sources
Field measurements	Catchment water balance	Easy to measure; low cost	Only long-term average is reliable	Sun *et al.* 2002
	Sap flow	Allows routine unsupervised measurement accurately at single plant scale	Large-scale measurement errors are determined by the sample size and variability of samples	Domec *et al.* 2012, Ford *et al.* 2007
	Eddy-covariance	Measures fluxes continuously, offering high temporal resolution data	High cost in instrumentation; gap filling required; energy imbalance problems	Baldocchi *et al.* 1988, Sun *et al.* 2008b
	Bowen ratio	Low cost; works for both crops and natural vegetation	Relies on several assumptions; errors associated with low gradients	Irmak *et al.* 2014, Bowen 1926
Remote sensing	Remote sensing	Spatially continuous; low temporal resolution	Uncertainties due to errors generated by measurement of sparse canopies; data mostly from clear sky conditions	Kustas and Norman 1996, Mu *et al.* 2007, Justice *et al.* 1998
Modeling	Theoretical models (*e.g.*, Penman-Monteith)	Widely used; long accepted; low cost	Requires site-specific parameters; not easy to apply on large scale	Penman 1948, Priestley and Taylor 1972, Allen *et al.* 1994
	Empirical (Budyko curves; flux data based)	Easy to understand; long-term mean estimate; easy to apply	May not be applicable to short-term estimates	Budyko *et al.* 1962, Zhang *et al.* 2004, Sun *et al.* 2011a

to environmental conditions (Domec *et al.* 2009). However, this method would be less reliable for forest stands with mixed species or where there is inappropriate sample size of measurement and structural scalars (Vinukollu *et al.* 2011). The eddy-covariance method measures fluxes continuously, offering high temporal resolution data series, but data availability is limited by costly site instrumentation and gap filling issues. In addition, the eddy-covariance method may underestimate ET by as much as 20% due to a lack of energy balance closure (Wilson *et al.* 2002). The Bowen ratio method estimates ET from the ratio of sensible heat and latent heat, using air temperature, humidity gradients, net radiation, and soil heat flux. It is relatively inexpensive but relies on several assumptions such as an extensive fetch over a homogeneous surface — a similar requirement for the eddy-covariance method (Heilman *et al.* 1989). Remote sensing has been widely used to estimate ET (Kustas and Norman 1996). This method has been regarded as a flexible technology to obtain large-scale ET and associated biophysical controls. MODIS global products (Mu *et al.* 2007) have provided spatially and temporally continuous ET estimates at a high resolution for modeling and analysis (Justice *et al.* 1998). However, estimation errors exist due to uncertainties in modeling effective surface emissivity and effective aerodynamic exchange resistance, and sparse canopies and thick clouds make remote sensing methods less reliable (Shuttleworth 2012). An example of estimated annual ET by biome using the eddy-covariance method is presented in Figure 4.1 to show the contrast of water uses in the USA.

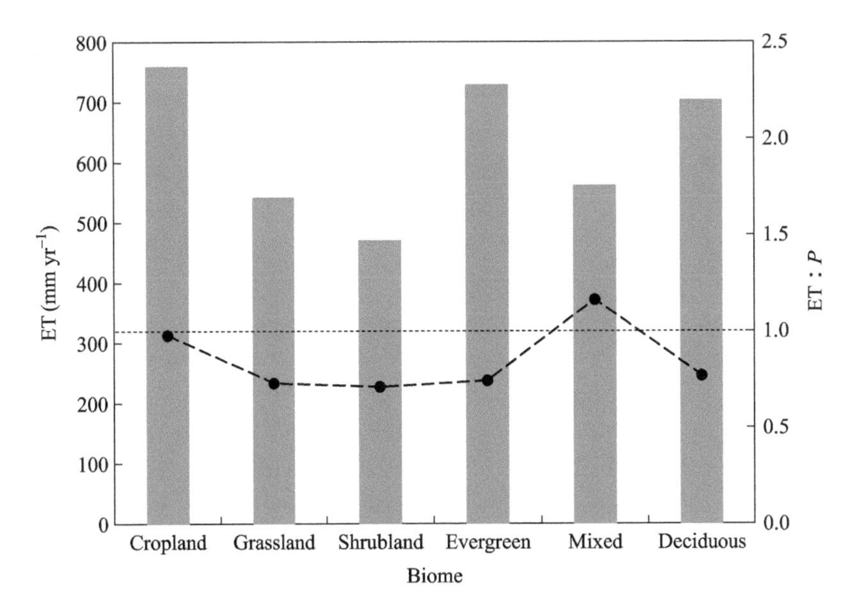

Fig. 4.1 Mean annual evapotranspiration (ET) and ET : precipitation (P) ratio by biome type measured using the eddy-covariance method from 74 sites within the AmeriFlux network (dashed line represents ET : P with reference to the right y axis) (Fang *et al.* 2016).

Due to the high cost of measuring ET at large scales, mathematical modeling has been widely used to estimate ET (McMahon *et al.* 2013). The primary ET models used in water balance and terrestrial ecosystem modeling include the theoretical methods (Penman 1948), Monteith (1965), and the Penman-Monteith model (Jensen *et al.* 1990). However, process-based models are difficult to use in practice due to limitations in parameterization and available climate data. Empirical models, which require fewer environmental variables and are easy-to-use, are consequently more commonly used in hydrologic modeling, especially at large spatial scale (Fang *et al.* 2016). Hydrological models that are designed to quantify discharge at watershed outlets must compute ET flux accurately to match stream flow observations. However, ET fluxes predicted by hydrological models are rarely calibrated or validated at the watershed scale (Tian *et al.* 2015). As a result, there exist large uncertainties in the model parameters, which potentially may result in "right answers for the wrong reasons" (Tian *et al.* 2012).

4.3 ET Models

Biophysical models are simplifications of the real world. Therefore, ET models are mathematical expressions that describe the ET processes at the plant, ecosystem, or landscape scale. ET models can be roughly divided into two groups, biophysical (or theoretical) and empirical models. Biophysical models are developed based on physical and physiological principles describing energy and water transport in the soil-plant-atmosphere continuum (SPA). Many of these models have evolved around the famous Penman-Monteith model that represents most advanced ET model. The Penman-Monteith model estimates ET as a function of available energy, vapor-pressure deficit (VPD), air temperature, vapor pressure, aerodynamic resistance (a function of primarily wind speed, and plant-canopy height and roughness) (Chapter 1), and canopy resistance (a measure of resistance to vapor transport from plants). In contrast, empirical ET models are developed using observed ET data and biophysical variables of plant characteristics, soil moisture, and atmospheric conditions by developing regression equations. Empirical models do not intend to describe the processes of vaporization, but can provide reasonable estimates with limited environmental input information.

In practice, it is often rather difficult to parameterize the process-based ET models to estimate actual ET, which is influenced by many factors such as stomata conductance (Chapter 2). To simplify the calculations, the concept of potential evapotranspiration was introduced by Penman in the 1940s. For any ecosystem, PET represents the potential water loss when soil water is not limiting. Actual ET (ET_a) then can be scaled down from the hypothetical PET by limiting canopy conductance, soil moisture and other constraining variables. PET models are often embedded in hydrological models that can simulate soil moisture dynamics. Here we present several PET methods and discuss the strengths and limitations for each.

4.3.1 PET Models

Existing PET models can be classified into five groups: (1) water budget (Guitjens 1982), (2) mass-transfer (Harbeck 1962), (3) combination (Penman 1948), (4) radiation driven, and (5) temperature based. There are approximately 50 models available to estimate PET, with models giving inconsistent values due to the assumptions, input data, or design (*i.e.*, for specific climatic regions). Previous studies at multiple scales have suggested that PET methods may give significantly different results (Lu *et al.* 2003, McMahon *et al.* 2013). In this chapter, we describe selected commonly used PET models listed in Table 4.2.

Table 4.2 Climatic variables and parameters required by the widely practiced PET models.

Method	Temperature	Radiation	Humidity	Others
FAO Penman-Monteith	Daily mean	Net radiation derived from solar radiation and extraterrestrial radiation; sunshine hours	Daily mean	Wind speed
Thornthwaite (1948)	Daily mean			Daytime length; heat index
Hamon (1963)	Daily mean			Daytime length; latitude
Blaney-Criddle (1950s)	Daily mean			Daytime length
Hargreaves-Samani (1982)	Daily maximum and minimum	Extraterrestrial radiation		Latitude
Priestley-Taylor (1972)	Daily mean	Net radiation derived from solar radiation and extraterrestrial radiation		Calibration constant (1.26)
Turc (1961)	Daily mean	Solar radiation	Daily mean	
Makkink (1957)	Daily mean	Solar radiation		

4.3.1.1 Penman-Monteith Model: FAO Reference ET Model

Potential evapotranspiration (PET) is a nebulous term. It can evoke confusion because PET does not clearly specify what land surface it refers to. The term, reference ET (ET_o) has gradually been replacing PET as a standard way to represent the energy conditions for a particular region, making PET estimates comparable worldwide. Using the process-based Penman-Monteith ET equation, actual daily ET of a hypothetical well-watered grass as reference ET (ET_o) is estimated using the following equation, referred as the FAO 56 model (Allen *et al.* 1994):

$$ET_o = \frac{0.408 \cdot \Delta \cdot (R_n - G) + \gamma \cdot \frac{Cn}{T+27 \cdot 3} \cdot u_2 \cdot (e_s - e_a)}{\Delta + \gamma \cdot (1 + Cd \cdot \mu_2)} \tag{4.4}$$

where

ET_o = grass reference ET (mm)

Δ = slope of the saturation water vapor pressure at air temperature (T, kPa °C^{-1})

$$\Delta = 2503 \frac{e^{\frac{17.27 \cdot T}{T+237.3}}}{(T + 237.3)^2}$$

R_n = net radiation (MJ m^{-2})

G = soil heat flux (MJ m^{-2})

γ = psychrometric constant (kPa °C^{-1})

e_s = saturation vapor pressure (kPa)

e_a = actual vapor pressure (kPa)

μ_2 = wind speed (m s^{-1}) at 2 m height

Cn = numerator constant that that changes with reference surface and calculation time step (900 °C mm s^{-3} Mg^{-1} d^{-1} for 24 h time steps, and 37 °C mm s^{-3} Mg^{-1} d^{-1} for hourly time steps)

Cd = denominator constant that changes with reference surface and calculation steps (0.34 s m^{-1} for 24 h time steps, 0.24 s m^{-1} for hourly time steps during daytime, and 0.96 s m^{-1} for hourly nighttime for grass reference surface) (Djaman *et al.* 2018).

This model assumes a stand that has a 0.12 m canopy height, a leaf area index (LAI) of 4.8, a bulk surface resistance of 70 s m^{-1}, and an albedo of 0.23. Details of the computation procedures can be found in Allen *et al.* (1994). Once ET_o is calculated, actual ET for the ecosystem can be estimated by simply multiplying a scalar K_c (crop coefficient). Crop coefficient is an empirical parameter that may vary by vegetation type, season, and disturbances (Allen *et al.* 1994).

As the most widely used model, the Penman-Monteith model has been presented in many different forms. These forms, nonetheless, all contain three key components (Allen *et al.* 1998). The first one is that $(R_n - G)$ represents available energy; the second one is the influences of vapor pressure deficit (*e.g.*, $e_s - e_a$); and the third one is an expression of canopy and atmospheric conductance that is related to canopy roughness and wind speed (*e.g.*, u^*; Chapters 1 and 2). Due to its broad uses, various numerical tools have also been developed (*e.g.*, Zotarelli *et al.* 2010).

4.3.1.2 Thornthwaite Model

The Thornthwaite (1948) PET model is the most widely used temperature-based monthly scale PET model because of its simplicity (*see also* Eq. 4.1). This model was derived by correlating mean monthly temperature with PET as determined from water balance for areas where sufficient moisture was available to maintain active transpiration.

$$\mathrm{PET} = 1.6 \cdot L_\mathrm{d} \cdot \left(\frac{10 \cdot T}{I}\right)^a \tag{4.5}$$

where

PET = monthly PET (cm)

L_d= mean daytime length (h), it is time from sunrise to sunset in multiples of 12 hours

T = monthly mean air temperature (°C)

$a = 6.75 \times 10^{-7} \cdot I^3 - 7.71 \times 10^{-5} \cdot I^2 + 0.01792 \cdot I + 0.49239$

I = annual heat index, which is computed from the monthly heat indices

$I = \sum_{j=1}^{12} i_j$

where

$i_j = \left(\frac{T_j}{5}\right)^{1.514}$

T_j= mean air temperature (°C) for month j; j = 1, ..., 12.

4.3.1.3 Hamon's PET Model

Hamon's PET model is also a temperature-based model (Hamon 1963). This method computes daily ET based on air temperature and theoretical daytime length (DAY). This model has been widely used in modeling studies on the impacts of climate change on water resources (Lu *et al.* 2005).

$$\mathrm{PET} = 0.1651 \cdot \mathrm{DAY} \cdot \frac{216.7 \cdot e_s}{t_a + 273.3} \quad (4.6)$$

where

$e_s = 6.108 \cdot e^{\frac{17.2694 \cdot t_a}{t_a + 237.3}}$

$\mathrm{DAY} = 2 \times \mathrm{acos}(-1 \times \tan(\mathrm{Lat} \times 0.0175) \times \tan\left(0.4093 \times \sin\left(\frac{2 \times 3.1415 \times \mathrm{DOY}}{365.0}\right) - 1.405\right)) / 3.14159$

where

PET = daily potential evapotranspiration (mm)

DAY = day length in multiples of 12 hours calculated from latitude and Julian day

e_s = saturation vapor pressure at a given temperature (mb)

t_a = mean air temperature (°C)

DOY = Julian day of the year ranging between 1 and 366

Lat = the latitude of the site

4.3.1.4 Blaney-Criddle PET Model

Blaney and Criddle (1957) proposed a model for estimating ET for the western USA and was modified by modified by Doorenbos and Pruitt (1977). The Blaney-Criddle equation has the following form:

$$\mathrm{PET} = P \cdot (0.46 \cdot T + 8.13) \quad (4.7)$$

where PET (mm) is the potential water use for a reference crop, T (°C) is mean temperature, and P (%) is percentage of total daytime hours for the period used (daily or monthly) out of total daytime hours of the year (365×12 = 4380 h). ET can be estimated as:

$$\mathrm{ET} = \mathrm{PET} \cdot k$$

where k is a monthly consumptive use coefficient, depending on vegetation type, location and season. According to Blaney (1959) for the growing season (May–October) k varies from 0.5 for orange tree to 1.2 for dense vegetation.

4.3.1.5 Turc PET Model

Turc (1961) simplified earlier versions of a PET (mm d^{-1}) equation for 10-day periods under general climatic conditions of Western Europe.

When relative humidity (RH) is < 50%

$$\text{PET} = 0.013\left(\frac{T}{T+15}\right)(R_s + 50)\left(1 + \frac{50 - \text{RH}}{70}\right) \tag{4.8}$$

when RH is > 50%

$$\text{PET} = 0.013\left(\frac{T}{T+15}\right)(R_s + 50)$$

where

T = daily mean air temperature (°C)

R_s= daily solar radiation (ly d^{-1}, or cal cm^{-2} d^{-1}); 1 cal cm^{-2} d^{-1} = (100/4.1868) (MJ m^{-2} d^{-1})

RH = daily mean relative humidity in percentage (%).

4.3.1.6 Priestley-Taylor Model

The Priestley-Taylor PET model (Priestley and Taylor 1972) was developed as a substitute to the Penman-Monteith equation to estimate ET when there is no soil water stress. For Priestley-Taylor model, only radiation observations are required. This is done by removing the aerodynamic terms from the Penman-Monteith equation and adding an empirically derived constant factor, α, of 1.26, when the general surrounding areas are wet or under humid conditions. In dry regions or seasons, α values are much higher than 1.26.

$$\lambda\text{PET} = \alpha\frac{\Delta}{\Delta + \gamma}(R_n - G) \tag{4.9}$$

where

PET = daily PET (mm d^{-1})

λ = latent heat of vaporization (MJ kg^{-1})

$$\lambda = 2.501 - 0.002361 \cdot T$$

T = daily mean air temperature(°C)

α = calibration constant, α = 1.26 for wet or humid conditions

Δ = slope of the saturation vapor pressure-temperature curve (kPa $°C^{-1}$)

$$\Delta = 0.200\ (0.00738 \cdot T + 0.8072)^7 - 0.000116$$

γ = psychrometric constant modified by the ratio of canopy resistance to atmospheric resistance (kPa $°C^{-1}$)

$$\gamma = \frac{c_p P}{0.622 \cdot \lambda}$$

c_p= specific heat of moist air at constant pressure (kJ kg^{-1} °C^{-1})

$$c_p = 1.013 \text{ kJ kg}^{-1} \,^\circ\text{C}^{-1} = 0.001013 \text{ MJ kg}^{-1} \,^\circ\text{C}^{-1}$$

P = atmospheric pressure (kPa)

$$P = 101.3 - 0.01055 \cdot \text{EL}$$

EL = elevation (m)
G = heat flux density to the ground (MJ m^{-2} d^{-1})

$$G = 4.2 \frac{T_{i+1} - T_{i-1}}{\Delta t} = -4.2 \frac{T_{i-1} - T_{i+1}}{\Delta t}$$

where
T_i = mean air temperature(°C) for the period i
ΔT= difference of time in days between two periods
R_n = net radiation (MJ m^{-2} d^{-1}). It is calculated as
$R_n = 0.77R_s - 2.45 \times 10^{-9} f \times [0.261 \times \exp(-7.7710 \times 10^{-4} T^2) - 0.02] \times (T_{max}^4 + T_{min}^4) + 0.83$
$f = \left(1.2 \times \frac{R_s}{R_a} + 0.1\right)$
where
R_s = solar radiation (MJ m^{-2} d^{-1})
R_a = extraterrestrial solar radiation (MJ m^{-2} d^{-1})
T = mean air temperature (K)
T_{max} = maximum air temperature (K)
T_{min}= minimum air temperature (K)

4.3.1.7 Makkink PET Model

Makkink (1957) estimated PET (mm d^{-1}) over 10-day periods for grassed lands under cool climatic conditions in the Netherlands as (Xu and Singh 2002):

$$\text{PET} = 0.61 \left(\frac{\Delta}{\Delta + \gamma}\right) \frac{R_s}{58.5} - 0.12 \qquad (4.10)$$

All variables in this equation have the same meanings and units as those in the Priestley-Taylor model (Eq. 4.9).

4.3.1.8 Hargreaves-Samani PET Model

The Hargreaves–Samani PET model was derived from eight years of cool season Alta fescue grass lysimeter data in Davis, California (Hargreaves and Samani 1982). There exist several equations for calculating PET, including the following one:

$$\lambda \cdot \mathrm{PET} = 0.0023 \cdot R_a \cdot \mathrm{TD}^{0.5} \cdot (T + 17.8) \tag{4.11}$$

where
- PET = daily PET (mm d^{-1})
- λ = latent heat of vaporization (MJ kg^{-1})
- T = daily mean air temperature (°C)
- R_a = extraterrestrial solar radiation (MJ m^{-2} d^{-1})
- TD = daily difference between the maximum and minimum air temperature (°C)

This method was later updated for reference ET (ET_o) estimates (Hargreaves and Samani 1985).

The Penman-Monteith, Priestley-Taylor, Hargreaves, and Stanghellini models (Stanghellini 1987) have all been widely used. A software developed by Donatelli *et al.* (2006) is useful one to fit these models with *in-situ* input variables.

4.3.2 Empirical Actual ET Models

Using field data collected from 13 sites across a variety of ET methods, Sun *et al.* (2011a) developed an empirical model for estimating monthly ET as a function of LAI, ET_o (mm mo^{-1}), and precipitation (mm mo^{-1}).

$$\mathrm{ET} = 11.94 + 4.76 \cdot \mathrm{LAI} + \mathrm{ET_o} \cdot (0.032 \cdot \mathrm{LAI} + 0.0026 \cdot P + 0.15) \tag{4.12}$$

where ET_o is the FAO 56 reference ET (Eq. 4.4) and P is monthly precipitation.

Another form of the ET model uses Hamon's PET model instead of the more data-demanding FAO reference ET method (Sun *et al.* 2011b):

$$\mathrm{ET} = 0.174 \cdot P + 0.502 \cdot \mathrm{PET} + 5.31 \cdot \mathrm{LAI} + 0.0222 \cdot \mathrm{PET} \cdot \mathrm{LAI} \tag{4.13}$$

Following a similar concept, Fang *et al.* (2016) employed the 250 FLUXNET synthsis dataset to develop the following two types of monthly ET models that require different input variables.

$$\mathrm{ET} = 0.42 + 0.74 \cdot \mathrm{PET} - 2.73 \cdot \mathrm{VPD} + 0.10 \cdot R_{\mathrm{n}} \tag{4.14}$$

where PET is monthly potential ET (mm) calculated by the Hamon's method. VPD (Pa, kPa) is estimated from air temperature and relative humidity (Chapter 1). Since R_{n} is rarely available at the regional scale, another model that uses commonly available data was developed:

$$\mathrm{ET} = -4.79 + 0.75 \cdot \mathrm{PET} + 3.92 \cdot \mathrm{LAI} + 0.04 \cdot P \tag{4.15}$$

$$R^2 = 0.68, \mathrm{RMSE} = 18.1 \text{ mm mo}^{-1}$$

Fang *et al.* (2016) further developed a series of monthly scale ET models by land cover type (Tables 4.3 and 4.4). These models accommodate users with different levels of access to climate data.

Table 4.3 Type I models by land cover type developed using the three most significant variables. These models are appropriate at a monthly scale. RMSE = root mean square error, R^2= coefficient of determination, n = number of monthly samples.

Land cover type	Model	RMSE (mm mo^{-1})	R^2	n
Shrubland	$\mathrm{ET} = -4.59 + 13.02 \cdot \mathrm{LAI} + 0.10 \cdot R_{\mathrm{n}} + 0.11 \cdot P$	11.2	0.85	193
Cropland	$\mathrm{ET} = 0.87 + 0.19 \cdot R_{\mathrm{n}} + 13.99 \cdot \mathrm{LAI} + 0.06 \cdot P$	20.2	0.72	649
Grassland	$\mathrm{ET} = -0.87 + 0.20 \cdot R_{\mathrm{n}} + 0.10 \cdot P + 0.24 \cdot \mathrm{SWC}$	15.7	0.73	562
Deciduous forest	$\mathrm{ET} = -14.22 + 0.74 \cdot \mathrm{PET} + 0.1 \cdot R_{\mathrm{n}}$	22.2	0.77	788
Evergreen needle leaf forest	$\mathrm{ET} = 13.47 + 0.10 \cdot R_{\mathrm{n}} + 1.35 \cdot T_{\mathrm{a}}$	17.2	0.71	1720
Evergreen broad leaf forest	$\mathrm{ET} = 0.01 + 0.63 \cdot T_{\mathrm{a}} + 0.46 \cdot \mathrm{SWC} + 0.14 \cdot R_{\mathrm{n}}$	12.5	0.90	69
Mixed forest	$\mathrm{ET} = -8.76 + 0.95 \cdot \mathrm{PET}$	13.1	0.79	259
Savannas	$\mathrm{ET} = -8.07 + 33.46 \cdot \mathrm{LAI} + 0.07 \cdot R_{\mathrm{n}}$	14.0	0.66	36

Units: ET, mm mo^{-1}; R_{n}, MJ mo^{-1}; P, mm mo^{-1}; PET, mm mo^{-1} estimated by Hamon's method; VPD, hPa; SWC, soil water content (%).

It is well known that long-term mean ET in a region is mainly controlled by water availability and atmosphere demand (Budyko *et al.* 1962). This relationship is well described in the Budyko-type of models (Zhang *et al.* 2004). Based on this concept, Zhang *et al.* (2001) analyzed watershed balances data for over 250 catchments worldwide and developed a simple two-parameter ET model that relates mean annual ET to rainfall, PET, and plant-available water capacity. The model offers a practical tool that can be readily used for assessing the long-term average effect of vegetation changes on catchment evapotranspiration.

Table 4.4 Type II models by land cover type developed using three commonly measured biophysical variables. RMSE = root mean square error, R^2 = coefficient of determination, n = number of monthly samples.

Land cover type	Model	RMSE (mm mo^{-1})	R^2	n
Shrubland	$\mathrm{ET} = -3.11 + 0.39 \cdot \mathrm{PET} + 0.09 \cdot P + 11.127 \cdot \mathrm{LAI}$	12.5	0.80	193
Cropland	$\mathrm{ET} = -8.15 + 0.86 \cdot \mathrm{PET} + 0.01 \cdot P + 9.54 \cdot \mathrm{LAI}$	20.9	0.70	653
Grassland	$\mathrm{ET} = -1.36 + 0.70 \cdot \mathrm{PET} + 0.04 \cdot P + 6.56 \cdot \mathrm{LAI}$	16.8	0.66	803
Deciduous forest	$\mathrm{ET} = -14.82 + 0.98 \cdot \mathrm{PET} + 2.72 \cdot \mathrm{LAI}$	23.7	0.74	754
Evergreen needle leaf forest	$\mathrm{ET} = 0.10 + 0.64 \cdot \mathrm{PET} + 0.04 \cdot P + 3.53 \cdot \mathrm{LAI}$	17.8	0.68	1382
Evergreen broad leaf forest	$\mathrm{ET} = 7.71 + 0.74 \cdot \mathrm{PET} + 1.85 \cdot \mathrm{LAI}$	16.8	0.76	233
Mixed forest	$\mathrm{ET} = -8.763 + 0.95 \cdot \mathrm{PET}$	13.1	0.79	259
Savannas	$\mathrm{ET} = -5.66 + 0.18 \cdot \mathrm{PET} + 0.10 \cdot P + 44.63 \cdot \mathrm{LAI}$	11.1	0.68	36

Units: ET, mm mo^{-1}; P, mm mo^{-1}; PET, mm mo^{-1} estimated by Hamon's method.

$$\mathrm{ET} = P \frac{1 + w \cdot \frac{\mathrm{PET}}{P}}{1 + w \cdot \frac{\mathrm{PET}}{P} + \frac{P}{\mathrm{PET}}} \tag{4.16}$$

where w is the plant-available water coefficient that represents the relative difference in plant water use for transpiration. PET as annual total can be estimated by Priestley-Taylor model. P is annual precipitation. The best fitted w values for forest and grassland are 2.0 and 0.5, respectively, when PET is estimated using the Priestley–Taylor model (Zhang *et al.* 2001). The w value can be as high as 2.8 when Hamon's PET model is applied for the humid southeastern USA (Sun *et al.* 2005).

By combining remote sensing and climate data for 299 river basins, Zeng *et al.* (2014) developed an annual ET model:

$$\mathrm{ET} = 0.4(\pm 0.02) \cdot P + 10.62(\pm 0.39) \cdot T + 9.63(\pm 2.27) \cdot \mathrm{NDVI} + 31.58(\pm 7.89), R^2 = 0.85 \tag{4.17}$$

where, ET is basin-averaged annual evapotranspiration (mm yr^{-1}), P, T and NDVI are annual precipitation (mm yr^{-1}), mean annual temperature (°C) and the annual average normalized difference vegetation index, respectively. This regression explains more than 85% of the spatiotemporal differences in ET across the 299 river basin years.

4.4 Model Demonstrations

4.4.1 Meteorological Data

Depending on methods used for estimating potential ET and actual ET, data requirements vary substantially. Here we use an example dataset to demonstrate the applications of major PET and actual ET models at different scales (hourly, daily, and monthly, and annual). Required microclimatic and biophysical variables, such as net radiation (R_n, W m^{-2}), air temperature (°C), wind speed (u, m s^{-1}), soil heat flux (G, W m^{-2}) and soil moisture (Ms, %) are presented to illustrate data requirements to use these models.

The one-year demonstration data sets (2016) were collected at one of the seven scale-up sites of the Great Lakes Bioenergy Research Center (GLBRC) at the Kellogg Biological Station (KBS) in southwestern Michigan, USA, with an open-path eddy-covariance flux tower (Zenone *et al.* 2011). This site (42°28′36.19″ N, 85°26′48.37″ W, 294 m a.s.l.) had been managed for more than 50 years as conventionally tilled corn-soybean crop field. The region lies on the northeastern edge of the US Corn Belt. The climate is temperate and humid, with a mean annual air temperature of 9.7 °C at KBS and an annual precipitation of 920 mm, evenly distributed throughout the year, with about half falling as snow. The soil textural class of all sites is sandy clay loam with a pH range from 5.8 to 6.4. The flux tower was installed in November 2008, and the mean fluxes of CO_2, H_2O and energy have been processed following conventional protocols of the FLUXNET (Abraha *et al.* 2019). Leaf area index (LAI) was not measured in 2016. An empirical linear model between LAI and NDVI from 2018 was used to back-predict LAI for the summer months of 2016. Day length (hour) for 2016 was calculated using the site-specific latitude and longitude with Solar. PY (S1-5) described in Chapter 1. Measured ET values at 30-min interval had a unit of W m^{-2} and were converted to mm at daily and monthly scales by excluding obvious outliers (Data4_1).

The mean annual ET for the site was estimated as 587 (±15) mm during 2010–2018, with a total precipitation of 951 mm and actual ET of 628 mm in 2016 (Abraha *et al.* 2019). Hourly ET can be as high as 0.35 mm per hour throughout the year. Except in July and August, hourly ET was normally less than 0.10 mm (Fig. 4.2a). Diel changes of ET are typical of those in the temperate zone (*i.e.*, low at night and high during the day), with peak

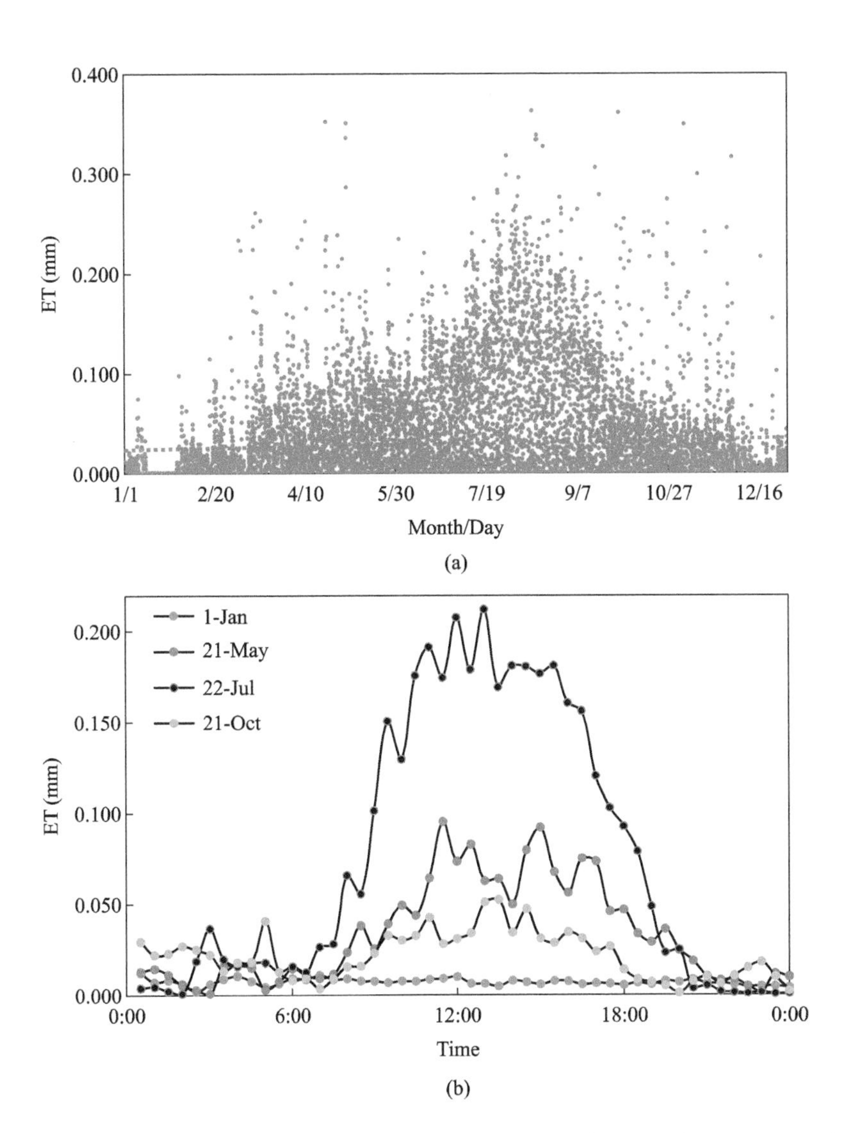

Fig. 4.2 (a) Change in hourly evapotranspiration (ET, mm) in 2016 and (b) the diel change on four selected days at a continuous corn site of the Kellogg Biological Station (KBS) in southwestern Michigan, USA, in 2016. Negative ET values from the flux tower were not shown.

hours during 10:00–16:00. The daily range was low in winter months and high during the summer (Fig. 4.2b). Calculated daily ET was low in the dormant season from October to April and high in the growing season, with five days higher than 5 mm d^{-1} in August (Fig. 4.3a). Monthly ET varied from 11 mm in January to 120 mm in August (Fig. 4.3b).

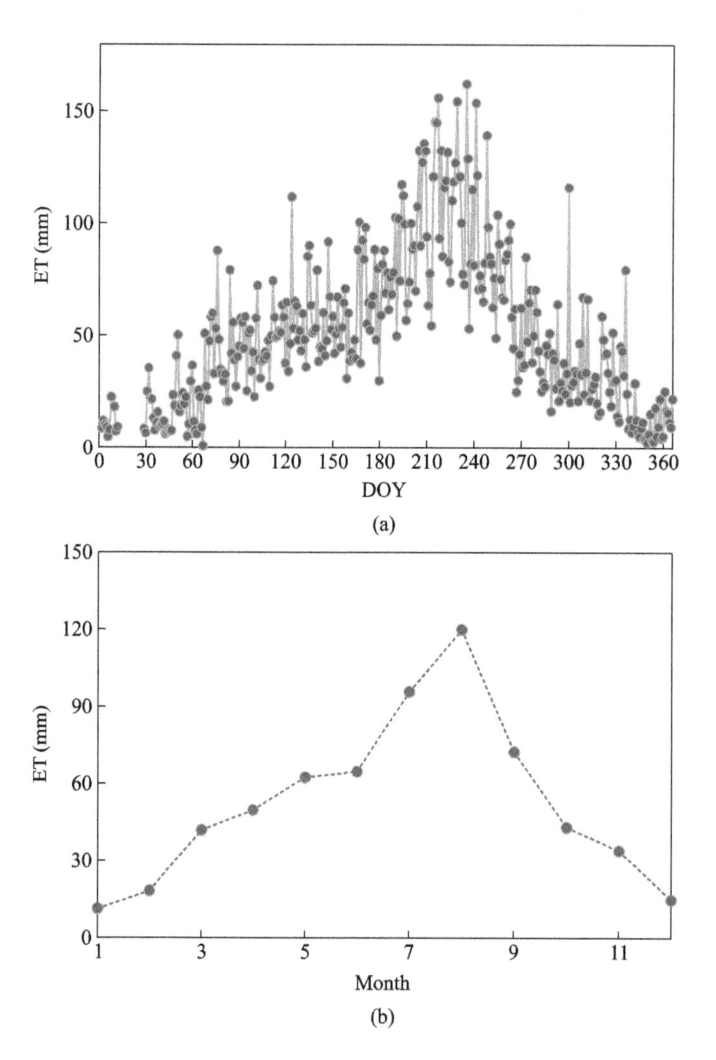

Fig. 4.3 Measured ET (mm) at (a) daily and (b) monthly scale using the eddy-covariance method at a continuous corn site of the Kellogg Biological Station (KBS) in southwestern Michigan, USA, in 2016. Negative ET values from the flux tower were not included in calculations.

4.4.2 Modeled PET at Multiple Scales and Actual ET

The FAO reference ET model (Eq. 4.4) is applied to estimate ET_o at a half-hour time intervals (Fig. 4.4) for the corn field at the Kellogg Biological Station (KBS) in southwestern Michigan, USA, in 2016. In this case, the 30-min meteorological variables and the specific parameters (*Cn, Cd*) for the hourly ET_o calculations are used. Similar to measured ET by the eddy-covariance method, ET_o rates are low in winter and highest in July and August (~0.46 mm per half hour).

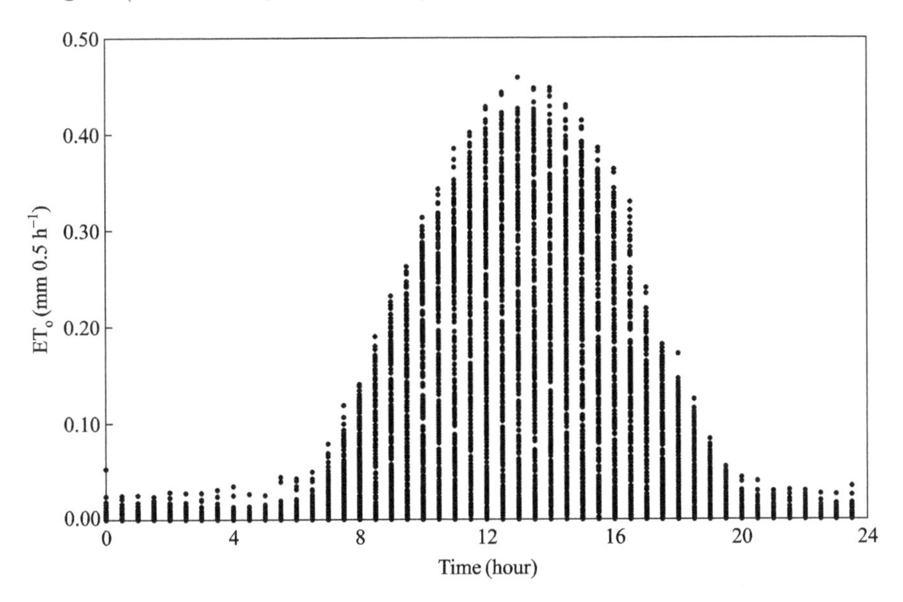

Fig. 4.4 Modeled 30-min reference evapotranspiration (ET_o) for a continuous corn field at the Kellogg Biological Station (KBS) in southwestern Michigan, USA, in 2016.

The FAO reference ET model (Eq. 4.4), the Priestley-Taylor equation (Eq. 4.9) representing the simplified Penman-Monteith model (Eq. 4.4), and the Hamon PET method (Eq. 4.6) representing a simple temperature-based PET model are applied to estimate ET_o at daily time interval (Fig. 4.5). Daily meteorological variables and the specific parameters (*Cn, Cd*) for daily ET_o calculations are used. Similar to measured daily ET by the eddy-covariance method, PET values are low in the winter and highest in July and August, but generally less than 10 mm per day. Hamon's method gives the lowest values among the three PET methods. As expected, the daily PET rates are generally higher than measured actual ET, most noticeably during the corn growing season when soil water stress is common at the site.

Two monthly-scale empirical ET models are applied to the same dataset

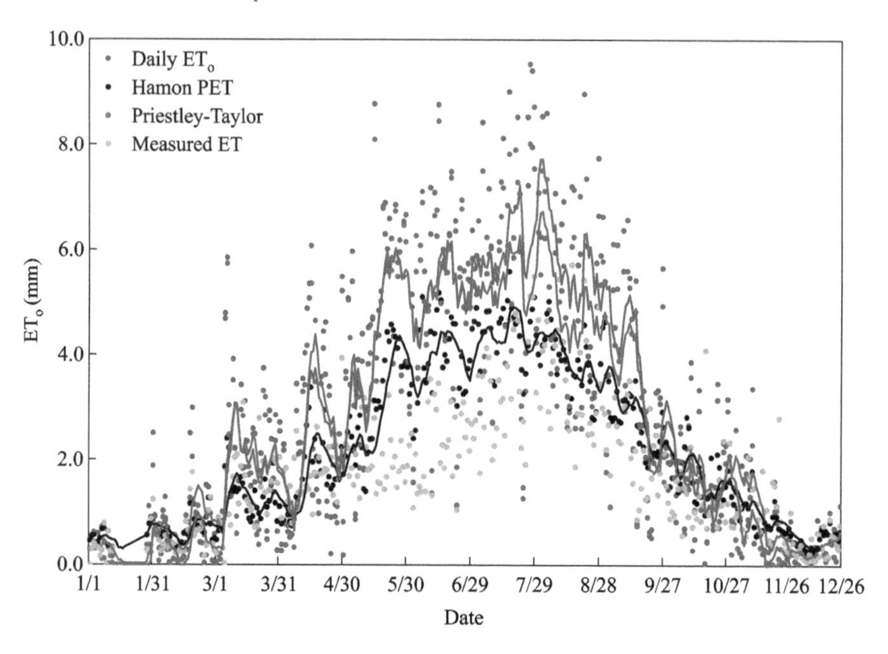

Fig. 4.5 Modeled daily reference ET (ET_o) and potential evapotranspiration (PET) with three biophysical models for a continuous corn field at the Kellogg Biological Station (KBS) in southwestern Michigan, USA, in 2016.

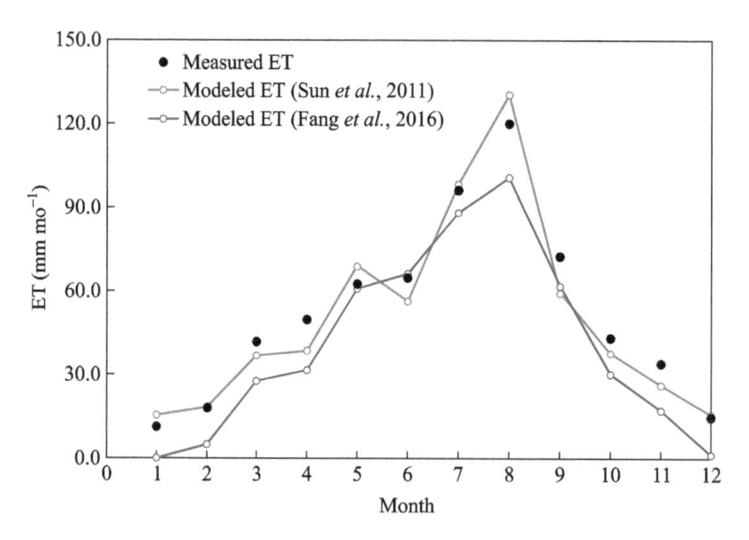

Fig. 4.6 Measured and modeled actual monthly evapotranspiration (ET, mm) using two empirical models for the continuous corn site at the Kellogg Biological Station (KBS) in southwestern Michigan, USA, in 2016.

for estimating actual ET for 2016 (Fig. 4.6). The two models published in Sun *et al.* (2011a) and Fang *et al.* (2016) represent a generalized ET model (Eq. 4.12) and a biome specific ET model (*i.e.*, the cropland model, Table 4.4), respectively. Both models require precipitation, PET, and LAI as model input. LAI was measured on site. Monthly PET calculated by Hamon's method was summarized from daily PET values. This exercise indicates that both models perform well to simulate the seasonal patterns and total annual ET (600–644 mm, compared to 628 mm from the measurements). It is clear that an ecosystem-specific model is not necessarily more accurate than a generalized model; there are always uncertainties associated with these empirical models.

Equation (4.16) may be used to estimate long-term annual ET for forests, grasslands, and other land cover types when P and PET are available. In this practice, $P = 920$ mm, PET $= 1000$ mm, and the empirical parameter w of 1.0 are used for demonstration purpose. The total annual ET is estimated to be 638 mm. This is considered to be high when compared to the long-term mean value reported for the study site (587±15 mm). It should be acknowledged that there is uncertainty to the w factor and it needs to be calibrated for local applications in practice.

$$\mathrm{ET} = 920\frac{1 + 1.0\frac{1000}{920}}{1 + 1.0\frac{1000}{920} + \frac{1000}{920}} = 638$$

4.5 Summary

The ET process serves as a major linkage among climatic, hydrologic, and ecological processes. Understanding the biophysical controls to ET helps understand other biological and physical processes of the Earth system. Thanks in part to the advances in micrometeorology and digital technology, much progress has been made in the past two decades towards measuring ET "everywhere all the time". However, the study of ET is still regarded as an imprecise science. Accurate quantification of water budgets, including water uses by ecosystems and humans, is becoming increasingly important given the growing competition for water resources among all users from agricultural irrigation and bioenergy development to domestic water withdrawals by cities in the Anthropocene. Climate change poses great environmental threats to ecosystems and water resources in the 21st century. Climate warming and the increase of the variability of precipitation form, amount, and timing can have ripple effects on ecosystem structure and functions through directly or indirectly altering the ET processes (*e.g.*, plant species change, water use efficiency). Similarly ET can change dramatically during land conversion from forests or wetlands to urban uses, resulting in urban environmental change such as increase in runoff, Urban Heat Island and Urban Dry Island effects. More reliable science on ET measurements must be developed to accurately scale it up or down among plot, watershed, regional, global scales to serve different purposes in natural resource management. In recent years remote sensing and radar technology have advanced rapidly and enhanced our capability to accurately quantify water use at a relatively fine scale. The best approach to estimate ET for practical applications is achieved by combining field measurements with high resolution remote sensing, and energy balance-based land surface modelling.

Online Supplementary Materials

S4-1: Field measurements of evapotranspiration (ET) and micrometeorological variables at 30 min interval in 2016 in an agricultural site (42°28′36.19″ N, 85°26′48.37″ W, 294 m a.s.l.) with an eddy-covariance tower of the Kellogg Biological Station, Michigan, USA (ETData.xlsx).
S4-2: Spreadsheet modeling of reference ET (ET_o), potential evapotranspiration (PET), and actual ET (Eqs. 4.4, 4.6, 4.12, 4.15, Table 4.4) for a corn field of the Kellogg Biological Station, Michigan, USA (ETModels.xlsx).

Scan the QR code or go to
https://msupress.org/supplement/BiophysicalModels
to access the supplementary materials.
User name: biophysical
Password: z4Y@sG3T

Acknowledgements

The authors appreciate Michael Abraha for providing the field data at a cropland of the Kellogg Biological Station. Ying-Ping Wang and Devendra Amatya provided insightful reviews on the chapter. Kristine Blakeslee edited the language and format of the draft.

References

Aber, J. D., and Federer, C. A. (1992). A generalized, lumped-parameter model of photosynthesis, evapotranspiration and net primary production in temperate and boreal forest ecosystems. *Oecologia*, 92(4), 463–474.

Abraha, M., Chen, J., Hamilton, S. K., and Robertson, G. P. (2019). Long-term evapotranspiration rates for rainfed corn versus perennial bioenergy crops in a mesic landscape. *Hydrological Processes*. 4, 810–822.

Allen, R. G., Smith, M., Perrier, A., and Pereira, L. S. (1994). An update for the definition of reference evapotranspiration. *International Commission on Irrigation and Drainage Bulletin*, 43(2), 1–34.

Allen, R. G., Pereira, L. S., Raes, D., and Smith, M. (1998). Chapter 2—FAO Penman-Monteith equation. *Crop Evapotranspiration—Guidelines for Computing Crop Water Requirements. FAO Irrigation and Drainage Paper*, 56. Rome.

Amatya, D., Sun, G., and Gowda, P. (2014). Evapotranspiration: Challenges in measurement and modeling: American Society of Agricultural and Biological Engineers (ASABE) International Symposium; Raleigh, North Carolina, 7–10 April 2014. *EOS Transactions American Geophysical Union*, 95(28), 256.

Baldocchi, D., Kelliher, F. M., Black, T. A., and Jarvis, P. (2000). Climate and vegetation controls on boreal zone energy exchange. *Global Change Biology*, 6(S1), 69–83.

Baldocchi, D. D., Hincks, B. B., and Meyers, T. P. (1988). Measuring biosphere-atmosphere exchanges of biologically related gases with micrometeorological methods. *Ecology*, 69(5), 1331–1340.

Baldocchi, D. D., and Ryu, Y. (2011). A synthesis of forest evaporation fluxes—from days to years—as measured with eddy covariance. In *Forest Hydrology and Biogeochemistry* (pp. 101–116). Springer, Dordrecht.

Blaney, H. F. (1959). Monthly consumptive use requirements for irrigated crops. *Journal of the Irrigation and Drainage Division*, 85(1), 1–12.

Blaney, H. F., and Criddle, W. D. (1957). Report on irrigation water requirements for Pakistan. *Bull IDFCR Council*, 3(2).

Bonan, G. B. (2008). Forests and climate change: Forcings, feedbacks, and the climate benefits of forests. *Science*, 320(5882), 1444–1449.

Bosch, J. M., and Hewlett, J. D. (1982). A review of catchment experiments to determine the effect of vegetation changes on water yield and evapotranspiration. *Journal of Hydrology*, 55(1–4), 3–23.

Bowen, I. S. (1926). The ratio of heat losses by conduction and by evaporation from any water surface. *Physical Review*, 27(6), 779.

Budyko, M. I., Yefimova, N. A., Aubenok, L. I., and Strokina, L. A. (1962). The heat balance of the surface of the earth. *Soviet Geography*, 3(5), 3–16.

Caldwell, P. V., Sun, G., McNulty, S. G., Cohen, E. C., and Myers, J. M. (2012). Impacts of impervious cover, water withdrawals, and climate change on river flows in the conterminous US. *Hydrology and Earth System Sciences*, 16, 2839–2857.

Campbell, G. S., and Norman, J. (2012). *An Introduction to Environmental Biophysics*. Springer Science and Business Media, 286 pp.

Chen, J., Ustin, S. L., Suchanek, T. H., Bond, B. J., Brosofske, K. D., and Falk, M. (2004). Net ecosystem exchanges of carbon, water, and energy in young and old-growth Douglas-fir forests. *Ecosystems*, 7(5), 534–544.

Currie, D. J. (1991). Energy and large-scale patterns of animal- and plant-species richness. *The American Naturalist*, 137(1), 27–49.

Domec, J. C., Noormets, A., King, J. S., Sun, G. E., McNulty, S. G., Gavazzi, M. J., Boggs, J. L., and Treasure, E. A. (2009). Decoupling the influence of leaf and root hydraulic conductances on stomatal conductance and its sensitivity to vapour pressure deficit as soil dries in a drained loblolly pine plantation. *Plant, Cell and Environment*, 32(8), 980–991.

Domec, J. C., Sun, G., Noormets, A., Gavazzi, M. J., Treasure, E. A., Cohen, E., Swenson, J. J., McNulty, S. G., and King, J. S. (2012). A comparison of three methods to estimate evapotranspiration in two contrasting loblolly pine plantations: Age-related changes in water use and drought sensitivity of evapotranspiration components. *Forest Science*, 58(5), 497–512.

Donatelli, M., Bellocchi, G., and Carlini, L. (2006). Sharing knowledge via software components: Models on reference evapotranspiration. *European Journal of Agronomy*, 24(2), 186–192.

Doorenbos, J., and Pruitt, W. O. (1977). *Guidelines for Predicting Crop Water Requirements. FAO Irrigation and Drainage Paper, 24* (revised). Rome.

Djaman, K., Koudahe, K., Lombard, K., and O'Neill, M. (2018). Sum of hourly vs daily Penman–Monteith grass-reference evapotranspiration under semiarid and arid climate. *Irrigation and Drainage Engineering*, 7(1), doi: 10.4172/2168-9768.1000202.

Fang, Y., Sun, G., Caldwell, P., McNulty, S. G., Noormets, A., Domec, J. C., King, J., Zhang, Z., Zhang, X., Lin, G., Zhou, G., Xiao, J., Chen, J., and Zhou, G. (2016). Monthly land cover-specific evapotranspiration models derived from global eddy flux measurements and remote sensing data. *Ecohydrology*, 9(2), 248–266.

Ford, C. R., Hubbard, R. M., Kloeppel, B. D., and Vose, J. M. (2007). A comparison of sap flux-based evapotranspiration estimates with catchment-scale water balance. *Agricultural and Forest Meteorology*, 145(3–4), 176–185.

Gedney, N., Cox, P. M., Betts, R. A., Boucher, O., Huntingford, C., and Stott, P. A. (2006). Detection of a direct carbon dioxide effect in continental river runoff records. *Nature*, 439(7078), 835–838.

Guitjens, J. C. (1982). Models of alfalfa yield and evapotranspiration. *Journal of the Irrigation and Drainage Division*, 108(3), 212–222.

Hamon, W. R. (1963). Computation of direct runoff amounts from storm rainfall. *International Association of Scientific Hydrology Publication*, 63, 52–62.

Hao, L., Huang, X., Qin, M., Liu, Y., Li, W., and Sun, G. (2018). Ecohydrological processes explain urban dry island effects in a wet region, southern China. *Water Resources Research*, 54(9), 6757–6771.

Harbeck, G. E. (1962). *A Practical Field Technique for Measuring Reservoir Evaporation Utilizing Mass-transfer Theory* (Vol. 272). US Government Printing Office. http://doi.org/10.3133/pp272E.

Hargreaves, G. H., and Samani, Z. A. (1982). Estimating potential evapotranspiration. *Journal of the Irrigation and Drainage Division*, 108(3), 225–230.

Hargreaves, G. H., and Samani, Z. A. (1985). Reference crop evapotranspiration from temperature. *Applied Engineering in Agriculture*, 1(2), 96–99.

Heilman, J. L., Brittin, C. L., and Neale, C. M. U. (1989). Fetch requirements for Bowen ratio measurements of latent and sensible heat fluxes. *Agricultural and Forest Meteorology*, 44(3–4), 261–273.

Hewlett, J. D. (1982). *Principles of Forest Hydrology.* University of Georgia Press, Athens, GA. 192 pp.

Irmak, S., Skaggs, K. E., and Chatterjee, S. (2014). A review of the Bowen ratio surface energy balance method for quantifying evapotranspiration and other energy fluxes. *Transactions of the ASABE*, 57(6), 1657–1674.

Jasechko, S., Sharp, Z. D., Gibson, J. J., Birks, S. J., Yi, Y., and Fawcett, P. J. (2013). Terrestrial water fluxes dominated by transpiration, *Nature*, 496(7445), 347–350.

Jensen, M. E., Burman, R. D. and Allen, R. G. (1990). Evaporation and irrigation water requirements. In *ASCE Manuals and Reports on Eng. Practices No. 70.* American Society of Civil Engineers, New York. 360 pp.

Jung, M., Reichstein, M., Ciais, P., Seneviratne, S. I., Sheffield, J., Goulden, M. L., Bonan, G., Cescatti, A., Chen, J., de Jeu, R., Dolman, A. J., Eugster, W., Gerten, D., Gianelle, D., Gobron, N., Heinke, J., Kimball, J., Law, B. E., Montagnani, L., Mu, Q., Mueller, B., Oleson, K., Papale, D., Richardson, A., Roupsard, O., Running, S., Tomelleri, E., Viovy, N., Weber, U., Williams, C., Wood, E., Zaehle, S., and Zhang, K. (2010). Recent decline in the global land evapotranspiration trend due to limited moisture supply. *Nature*, 467(7318), 951–954.

Justice, C. O., Vermote, E., Townshend, J. R., Defries, R., Roy, D. P., Hall, D. K., Salomonson, V. V., Privette, J. L., Riggs, G., Strahler, A., Lucht, W., Myneni, R. B., Knyazikhin, Y., Running, S. W., Nemani, R. R., Wan Huete, A. R., van Leeuwen, W., Wolfe, R. E., Giglio, L., Muller, J.-P., Lewis, P., and Barnsley, M. J. (1998). The Moderate Resolution Imaging Spectroradiometer (MODIS): Land remote sensing for global change research. *IEEE Transactions on Geoscience and Remote Sensing*, 36(4), 1228–1249.

Koster, R. D., Dirmeyer, P. A., Guo, Z., Bonan, G., Chan, E., Cox, P., Gordon, C. T., Kanae, S., Kowalczyk, E., Lawrence, D., Liu, P., Lu, C-H., Malyshev, S., McAvaney, B., Mitchell, K., Mocko, D., Oki, T., Oleson, K., Pitman, A., Sud, Y. C., Taylor, C. M., Verseghy, D., Vasic, R., Xue, Y., and Yamada, T. (2004). Regions of strong coupling between soil moisture and precipitation. *Science*, 305(5687), 1138–1140.

Kustas, W. P., and Norman, J. M. (1996). Use of remote sensing for evapotranspiration monitoring over land surfaces. *Hydrological Sciences Journal*, 41(4), 495–516.

Lu, J., Sun, G., McNulty, S. G., and Amatya, D. M. (2003). Modeling actual evapotranspiration from forested watersheds across the southeastern United States. *Journal of the American Water Resources Association*, 39(4), 886–896.

Lu, J., Sun, G., McNulty, S. G., and Amatya, D. M. (2005). A comparison of six potential evapotranspiration methods for regional use in the southeastern United States. *Journal of the American Water Resources Association*, 41(3), 621–633.

Makkink, G. F. (1957). Testing the Penman formula by means of lysimeters. *Journal of the Institution of Water Engineerrs*, 11, 277–288.

McMahon, T. A., Peel, M. C., Lowe, L., Srikanthan, R., and McVicar, T. R. (2013). Estimating actual, potential, reference crop and pan evaporation using standard meteorological data: A pragmatic synthesis. *Hydrology and Earth System Sciences*, 17(4), 1331–1363.

Monteith, J. L. (1965). Evaporation and environment. pp. 205–234. In Fogg, G. E. *Symposium of the Society for Experimental Biology, The State and Movement of Water in Living Organisms,* Vol. 19. Academic Press, Inc., NY.

Mu, Q., Heinsch, F. A., Zhao, M., and Running, S. W. (2007). Development of a global evapotranspiration algorithm based on MODIS and global meteorology data. *Remote Sensing of Environment*, 111(4), 519–536.

Oki, T., and Kanae, S. (2006). Global hydrological cycles and world water resources. *Science*, 313(5790), 1068–1072.

Oudin, L., Andréassian, V., Lerat, J., and Michel, C. (2008). Has land cover a significant impact on mean annual streamflow? An international assessment using 1508 catchments. *Journal of Hydrology*, 357(3–4), 303–316.

Peng, S. S., Piao, S., Zeng, Z., Ciais, P., Zhou, L., Li, L. Z., Myneni, R. B., and Zeng, H. (2014). Afforestation in China cools local land surface temperature. *Proceedings of the National Academy of Sciences*, 111(8), 2915–2919.

Penman, H. L. (1948). Natural evaporation from open water, bare soil and grass. *Proceedings of the Royal Society of London. Series A. Mathematical and Physical Sciences*, 193(1032), 120–145.

Penman, H. L. (1963). Vegetation and hydrology. *Soil Science*, 96(5), 357.

Priestley, C. H. B., and Taylor, R. J. (1972). On the assessment of surface heat flux and evaporation using large-scale parameters. *Monthly Weather Review*, 100(2), 81–92.

Sanford, W. E., and Selnick, D. L. (2013). Estimation of evapotranspiration across the conterminous United States using a regression with climate and land-cover data 1. *JAWRA Journal of the American Water Resources Association*, 49(1), 217–230.

Savenije, H. H. (2004). The importance of interception and why we should delete the term evapotranspiration from our vocabulary. *Hydrological Processes*, 18(8), 1507–1511.

Schellekens, J., Dutra, E., la Torre, A. M. D., Balsamo, G., van Dijk, A., Weiland, F. S., Minvielle, M., Calvet, J-C., Decharme, B., Eisner, S., Fink, G., Flörke, M., Peßenteiner, S., van Beek, R., Polcher, J., Beck, H., Orth, R., Calton, B., Burke, S., Dorigo, W., and Weedon, G. P. (2017). A global water resources ensemble of hydrological models: The eartH_2Observe Tier-1 dataset. *Earth System Science Data*, 9, 389–413.

Shuttleworth, W. J. (2012). *Terrestrial Hydrometeorology*. John Wiley and Sons. 480p.

Smith, D. M., and Allen, S. J. (1996). Measurement of sap flow in plant stems. *Journal of Experimental Botany*, 47(12), 1833–1844.

Stanghellini, C. (1987). *Transpiration of Greenhouse Crops:An Aid to Climate Management,*. Doctoral Dissertation, IMAG, Agricultural University, Weningen, The Netherlands. 150pp.

Stanhill, G. (2005). Evapotranspiration. *Encyclopedia of Soils in the Environment*, 502–506.

Stigter, E. E., Litt, M., Steiner, J. F., Bonekamp, P. N., Shea, J. M., Bierkens, M. F., and Immerzeel, W. W. (2018). The importance of snow sublimation on a Himalayan glacier. *Frontiers in Earth Science*, 6, 108.

Sun, G., Alstad, K., Chen, J., Chen, S., Ford, C. R., Lin, G., Liu, C., Lu, N., McNulty, S. G., Miao, H., Noormets, A., Vose, J. M., Wilske, B., Zeppel, M., Zhang, Y., and Zhang, Z. (2011a). A general predictive model for estimating monthly ecosystem evapotranspiration. *Ecohydrology*, 4(2), 245–255.

Sun, G., Caldwell, P., Noormets, A., McNulty, S. G., Cohen, E., Moore Myers, J. M., Domec, J-C., Treasure, E., Mu, Q., Xiao, J., John, R., and Chen, J. (2011b). Upscaling key ecosystem functions across the conterminous United States by a water-centric ecosystem model. *Journal of Geophysical Research: Biogeosciences*, 116(G3), https://doi.org/10.1029/2010JG001573.

Sun, G., McNulty, S. G., Amatya, D. M., Skaggs, R. W., Swift Jr, L. W., Shepard, J. P., and Riekerk, H. (2002). A comparison of the watershed hydrology of coastal forested wetlands and the mountainous uplands in the Southern US. *Journal of Hydrology*, 263(1–4), 92–104.

Sun, G., McNulty, S. G., Lu, J., Amatya, D. M., Liang, Y., and Kolka, R. K. (2005). Regional annual water yield from forest lands and its response to potential deforestation across the southeastern United States. *Journal of Hydrology*, 308(1–4), 258–268.

Sun, G., McNulty, S. G., Moore Myers, J. A., and Cohen, E. C. (2008a). Impacts of multiple stresses on water demand and supply across the southeastern United States. *Journal of the American Water Resources Association*, 44(6), 1441–1457.

Sun, G., Noormets, A., Chen, J., and McNulty, S. G. (2008b). Evapotranspiration estimates from eddy covariance towers and hydrologic modeling in managed forests in Northern Wisconsin, USA. *Agricultural and Forest Meteorology*, 148(2), 257–267.

Sun, G., Noormets, A., Gavazzi, M. J., McNulty, S. G., Chen, J., Domec, J. C., King, J. S., Amatya, D. M., and Skaggs, R. W. (2010). Energy and water balance of two contrasting loblolly pine plantations on the lower coastal plain of North Carolina, USA. *Forest Ecology and Management*, 259(7), 1299–1310.

Thornthwaite, C. W. (1948). An approach toward a Rational classification of climate. *Geographical Review*, 38(1), 55–94.

Tian, S., Youssef, M. A., Skaggs, R. W., Amatya, D. M., and Chescheir, G. M. (2012). Modeling water, carbon, and nitrogen dynamics for two drained pine plantations under intensive management practices. *Forest Ecology and Management*, 264, 20–36.

Tian, S., Youssef, M. A., Sun, G., Chescheir, G. M., Noormets, A., Amatya, D. M., Skaggs, R. W., King, J. S., McNulty, S., Gavazzi, M., Miao, G., and Domec, J-C. (2015). Testing DRAINMOD-FOREST for predicting evapotranspiration in a mid-rotation pine plantation. *Forest Ecology and Management*, 355, 37–47.

Trenberth, K. E., Fasullo, J. T., and Kiehl, J. (2009). Earth's global energy budget. *Bulletin of the American Meteorological Society*, 90(3), 311–324.

Turc, L. (1961). Estimation of irrigation water requirements, potential evapotranspiration: A simple climatic formula evolved up to date. *Annuals of Agronomy*, 12(1), 13–49.

Vinukollu, R. K., Wood, E. F., Ferguson, C. R., and Fisher, J. B. (2011). Global estimates of evapotranspiration for climate studies using multi-sensor remote sensing data: Evaluation of three process-based approaches. *Remote Sensing of Environment*, 115(3), 801–823.

Vose, J. M., Sun, G., Ford, C. R., Bredemeier, M., Otsuki, K., Wei, X., Zhang, Z., and Zhang, L. (2011). Forest ecohydrological research in the 21st century: What are the critical needs? *Ecohydrology*, 4(2), 146–158.

Webb, E. K. (1960). On estimating evaporation with fluctuating Bowen ratio. *Journal of Geophysical Research*, 65(10), 3415–3417.

Wilson, K., Goldstein, A., Falge, E., Aubinet, M., Baldocchi, D., Berbigier, P., Bernhofer, C., Ceulemans, R., Dolman, H., Field, C., Grelle, A., Ibrom, A., Law, B. E., Kowalski, A., Meyers, T., Moncrieff, J., Monson, R., Oechel, W., Tenhunen, J., Valentini, R., and Verma, S. (2002). Energy balance closure at FLUXNET sites. *Agricultural and Forest Meteorology*, 113(1–4), 223–243.

Wilson, K. B., Hanson, P. J., Mulholland, P. J., Baldocchi, D. D., and Wullschleger, S. D. (2001). A comparison of methods for determining forest evapotranspiration and its components: Sap-flow, soil water budget, eddy covariance and catchment water balance. *Agricultural and Forest Meteorology*, 106(2), 153–168.

Xu, C. Y., and Singh, V. P. (2002). Cross comparison of empirical equations for calculating potential evapotranspiration with data from Switzerland. *Water Resources Management*, 16(3), 197–219.

Zeng, Z., Wang, T., Zhou, F., Ciais, P., Mao, J., Shi, X., and Piao, S. (2014). A worldwide analysis of spatiotemporal changes in water balance-based evapotranspiration from 1982 to 2009. *Journal of Geophysical Research: Atmospheres*, 119(3), 1186–1202.

Zenone, T., Chen, J., Deal, M. W., Wilske, B., Jasrotia, P., Xu, J., Bhardwaj, A. K., Hamilton, S. K., and Robertson, G. P. (2011). CO_2 fluxes of transitional bioenergy crops: Effect of land conversion during the first year of cultivation. *Global Change Biology-Bioenergy*, 3(5), 401–412.

Zhang, L., Dawes, W. R., and Walker, G. R. (2001). Response of mean annual evapotranspiration to vegetation changes at catchment scale. *Water Resources Research*, 37(3), 701–708.

Zhang, L., Hickel, K., Dawes, W. R., Chiew, F. H., Western, A. W., and Briggs, P. R. (2004). A rational function approach for estimating mean annual evapotranspiration. *Water Resources Research*, 40(2), doi:10.1029/2003WR002710.

Zotarelli, L., Dukes, M. D., Romero, C. C., Migliaccio, K. W., and Morgan, K. T. (2010). *Step by Step Calculation of the Penman-Monteith Evapotranspiration (FAO-56 Method)*. Institute of Food and Agricultural Sciences. University of Florida. Gainesville, FL. 10pp.

Chapter 5
Modeling Ecosystem Global Warming Potentials

Jiquan Chen, Cheyenne Lei and Pietro Sciusco

5.1 Introduction

Scientific investigations on global warming potential (GWP) have seen escalating growth since 1990 amid a rapid global warming trend and its profound impacts on nature and society. To explore the surrounding issues and the needs for modeling GWP, here we will introduce the physics of Earth energy balance (Section 5.1), calculate GWP of three major greenhouse gas species and albedo (Sections 5.2–5.3), and provide demonstration examples (Section 5.4). GWP refers to the net effect of greenhouse gas (GHG) emissions as well as of the albedo changes on the radiative balance of the Earth and therefore

Jiquan Chen
Landscape Ecology & Ecosystem Science (LEES) Lab, Department of Geography, Environment, and Spatial Sciences & Center for Global Change and Earth Observations, Michigan State University, East Lansing, MI 48823
Email: jqchen@msu.edu

Cheyenne Lei
Landscape Ecology & Ecosystem Science (LEES) Lab, Department of Geography, Environment, and Spatial Sciences & Center for Global Change and Earth Observations, Michigan State University, East Lansing, MI 48823
Email: cheyenne@msu.edu

Pietro Sciusco
Landscape Ecology & Ecosystem Science (LEES) Lab, Department of Geography, Environment, and Spatial Sciences, Michigan State University, East Lansing, MI 48823
Email: sciuscop@msu.edu

Jiquan Chen, *Biophysical Models and Applications in Ecosystem Analysis*,
https://doi.org/10.3868/978-7-04-055256-0-5

its temperature warming potential over a specified time period, usually 100 years. Because the effect of a GHG is determined by both its physical interaction with radiation and its chemical life span in the atmosphere, GWP cannot easily be determined by first principles alone and instead are usually calculated by simulating the climate effect of an emission using energy balance and Earth system models. Typically, GWP are rendered unitless by being normalized against the GWP of a long-lived gas, usually CO_2. The purpose of this chapter is to provide algorithms for calculating relative and absolute GWP following IPCC protocols and to demonstrate the use of these calculations through a spreadsheet model that predicts GWP of three major GHGs and albedo.

5.1.1 Temperature of the Earth

Temperature within the Earth system is maintained by the energy balance—received from the sun and reflected/emitted from the Earth. Earth receives radiation from its cross section area (πr^2, *a.k.a.* shadow of the Earth) but emits radiation as a sphere (*i.e.*, $4\pi r^2$). The amount of solar radiation at the top of the atmosphere is a constant (342 W m^{-2}). When this is converted to a spherical surface, the actual amount of energy at the top of the atmosphere is four times this value at ~1370 W m^{-2} — the solar constant. Of the radiation that permeates the atmosphere, part is absorbed by the atmosphere, part of it reaches the ground, and approximately 30% is reflected back to outer space by clouds and the surface. The ratio between the reflection of solar radiation and the total incoming solar radiation is defined as albedo (α_s), with a range of 0–1. The remaining 70% of this radiation is absorbed by the Earth and emitted as longwave radiation. After accounting for absorption, the surface radiation is closer to 50% of incoming TOA (top of atmosphere) (Trenberth *et al.* 2009). The blackbody theory states that any object with a temperature higher than absolute zero (−273.15 °C) emits energy. According to Stefan-Boltzmann law, the total energy (W m^{-2}) radiated from Earth is proportional to the fourth power of Earth's temperature (T):

$$E = \varepsilon \cdot \sigma \cdot T^4 \tag{5.1}$$

where σ is the Stefan-Boltzmann constant (1.38054×10^{-23} J K^{-1}) and ε is the emissivity (W m^{-2} μm^{-1}) of the Earth. Emissivity varies by wavelength (λ, μm). With a stable atmosphere, the incoming radiation energy and outgoing emittance is balanced, resulting in an approximate Earth temperature of ~15 °C (~288 K). Notably without any greenhouse gases in the atmosphere, Earth's temperature would be 255 K (*i.e.*, atmospheric effective emissivity of 0).

Both solar radiation and emitted energy from Earth are transported through photons. These photons travel at the speed of light ($c = 3 \times 10^{10}$ m s^{-1}) and behave as particles and waves. The spectral density of radiation by wavelength under a thermal equilibrium at a given temperature (T) is calculated as:

$$E = \frac{h \cdot c}{\lambda} \tag{5.2}$$

where h is Planck's constant (6.6256×10^{-34} J s), and λ is the wavelength of the photons (μm). Planck's law also defines the radiant spectral flux density as:

$$E(\lambda, T) = \frac{2 \cdot \pi \cdot h \cdot c^2}{\lambda^5 \cdot \left(e^{\frac{h \cdot c}{\sigma \cdot \lambda \cdot T}} - 1\right)} \tag{5.3}$$

Equations (5.2) and (5.3) define the radiant emittance from the Earth. Based on Planck's law, radiant emittance from the sun ($T = 6000$ K) varies from 0.2 nm to 3.0 nm and peaks at spectra of ~0.55 nm, with most energy coming from a wavelength of 0.3–0.7 μm (>50%). These wavelengths are loosely called "shortwave radiation" or "visible light". The Earth's ($T = 288$ K) peak emittance is 9.5–10.0 nm. In sum, Earth's temperature is mostly maintained by the balance of incoming shortwave radiation from the sun and outgoing longwave radiation from the Earth.

5.1.2 The Greenhouse Effects

This energy balance is maintained by assuming that three major components of the Earth system remain the same: water (oceans and inland lakes), land surface, and the atmosphere. While the division between land and water has been stable since the last glaciation, atmospheric composition and land surface properties have gone through rapid and magnificent changes, mostly due to increasing use of fossil fuels and land conversions. Excessive consumption of fossil fuels (*e.g.*, coals, crude oil, natural gases, *etc.*) has been substantially changing atmospheric composition, which in turn alters both incoming shortwave and outgoing longwave radiation. Due to extensive and intensive land use, terrestrial ecosystems have very different capacity for converting atmospheric CO_2 to biomass through plant photosynthesis (Chapter 2), uptaking CH_4 into soils through methane oxidation, producing CH_4 in the soil, decomposing dead organic materials for respiring (Chapter 3), partitioning latent and sensible heat differently (Chapters 1 and 4), releasing more N_2O due to elevated fertilization, and others. All of these changes in ecosystem structure and functions have direct consequences on many chemical species of the atmosphere. As illustrated in Figure 5.1a, the CO_2 concentration of the atmosphere increased from 337 ppm in 1979 to 407 ppm in 2018. Similarly, N_2O and CH_4 concentrations that were 301 ppb and 1578 ppb, respectively, in 1979 rose to 331 ppb and 1858 ppb, respectively, in 2018. The decadal increase in CO_2, N_2O and CH_4, on average, was 17.6 ppm, 7.5 ppb, 59.2 ppb, respectively, during this 40-year period. As a result, the radiative forcing (RF) due to these increasing concentrations has been elevated, with a higher increasing rate for CO_2 and N_2O than CH_4 (Fig. 5.1b). When examined for their relative importance of the total, CO_2 increased from 66.8% to 74.2%, CH_4 decreased from 26.4% to 18.6%, and N_2O increased from 6.8% to 7.2% (Fig. 5.1c). These changes in their importance in contributing the warming trend highlight the recent increasing interest in N_2O.

Earth's atmosphere is composed of many gases and particles. Nitrogen (~78%) and oxygen (~21%) are the two major chemical species. Other species, including water, argon, carbon dioxide, neon, helium, methane, krypton, hydrogen, nitrous oxide, xenon, ozone, carbon monoxide, and ammonia, account for <1%. The atmosphere also contains a large amount of particulate matters and biological materials that can affect the energy balance of the Earth. For example, an increase in the amount of aerosol deposition on snow and glacier surfaces will reduce reflection (*i.e.*, lowered albedo) and, consequently, keep more energy within the Earth system. Volcanic eruptions, large-scale fires, and industrial production are major sources of aerosols at global scale.

If these gases and particular matters had the same reflection (or scattering) and absorption to radiation at all wavelengths, the energy balance of the Earth would not be affected. Unfortunately, this is not the case (Fig. 5.2). Ozone (O_3) absorbs almost all wavelengths of <300 nm (*i.e.*, ultraviolet radiation, UV), and water (H_2O) absorbs many wavelengths of >700 nm. This suggests that a reduction in ozone concentration would increase the amount of incoming UV light on the Earth's surface. This process witnessed rapid change during a period of increased chlorofluorocarbon emissions from a variety of industrial processes until their ban in the late 1980s that reduced ozone concentration in the stratosphere. Water vapor emissions from high-altitude flight also contribute to this ozone depletion, but this is

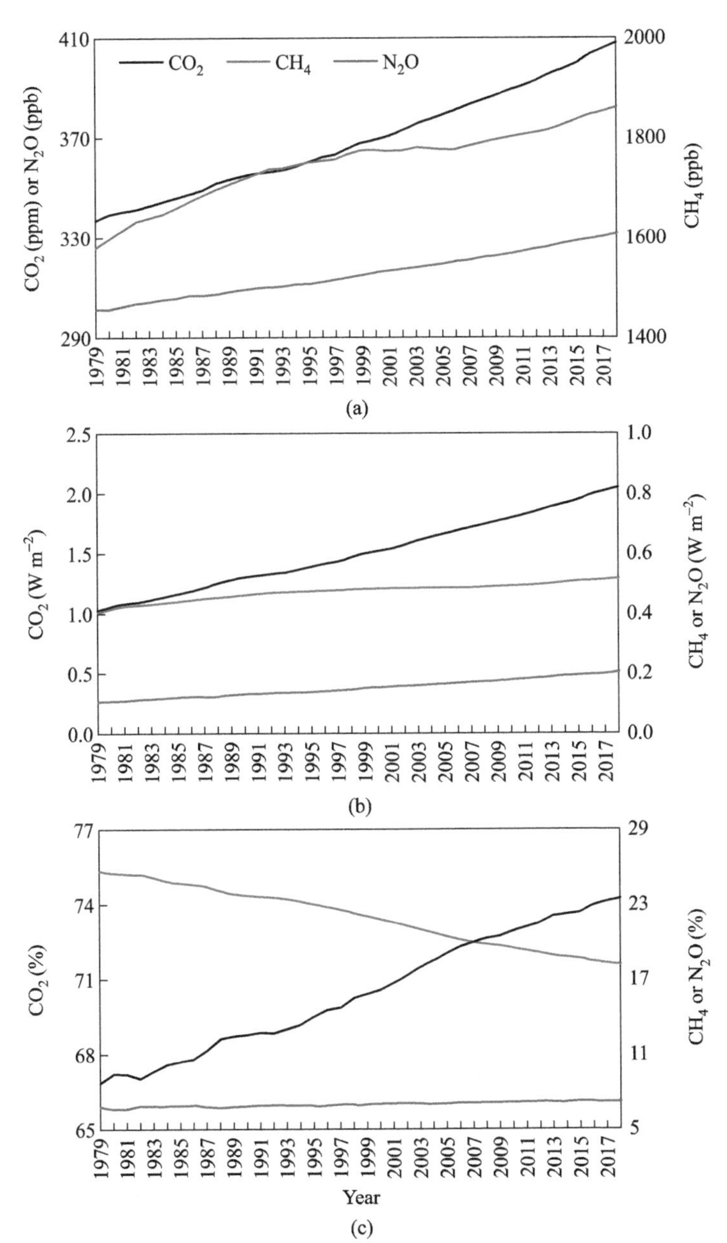

Fig. 5.1 Long-term changes in three major greenhouse gas (GHG) species (CO_2, CH_4 and N_2O) (a) concentrations, (b)their radiative forcing (RF), and (c) their RF portion of the total during 1979–2018 [1].

---

1 Data source: https://www.esrl.noaa.gov/gmd/aggi/NOAA_MoleFractions_2019.csv(downloaded on 29 March 2020).

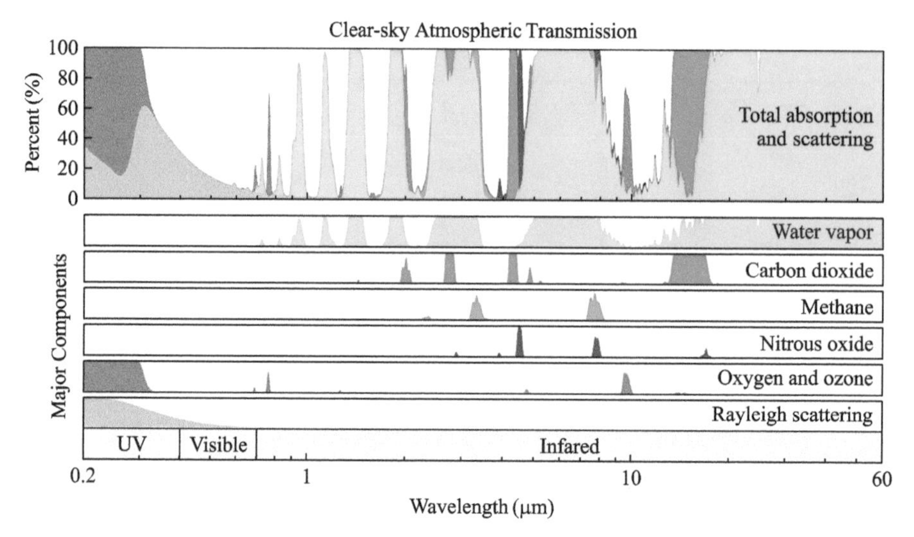

Fig. 5.2 Spectral distribution of solar radiation intensity and absorptions by water vapor and major greenhouse gases. Absorptions were computed using the HiTRAN database[1], and the US Standard Atmosphere with the assistance of Brad Schrag's PyRad analysis package by Robert Rhode of Berkeley Earth (permission received on April 3, 2020).

a much smaller cause of ozone loss than CFCs. Fortunately, according to Planck's law (Eq. 5.3), the total amount of energy from UV radiation is small because of the shorter wavelength. Meanwhile, water vapor can absorb a large amount of longwave radiation in a large number of wavelengths, and thus is the primary greenhouse gas on Earth, resulting in a planet that maintains a temperature that supports life. Liquid and frozen water behaves similarly. This is the reason cloudy nights are typically warmer than clear nights. Fortunately, water vapor has a short lifespan (*e.g.*, hours to days) in the atmosphere and its quantity in the atmosphere is determined by evaporation and condensation processes that are functions of temperature. Humans cannot significantly modify water vapor concentration. Rather, water vapor changes as temperature changes due to changes in greenhouse gases like CO_2. Thus the climate effect of change in water vapor is called a climate feedback. Roughly, the relative humidity of Earth has stayed constant.

Unlike H_2O, there are serious concerns about trace gases that humans can modify directly by emissions. As shown in Figure 5.2, GHGs such as CH_4, N_2O, O_2, and O_3 can absorb radiation at several wavelengths, with

1 https://hitran.org/data-index/

CO_2, CH_4 and N_2O representing the top three species in terms of absorption of longwave radiation. The abovementioned mechanism for keeping energy within a system is illustrated in pioneering experiments by John Tyndall and Joseph Fourier, whose work in the mid-1800s found that glass lets most shortwave radiation through, while it intercepts most longwave radiation — known as the "greenhouse effect". This principle is widely used to grow vegetables and crops in glass greenhouses because a greenhouse reduces turbulent mixing while allowing for solar absorption that leads to a higher temperature. In the case of the Earth system, the atmosphere functions similarly to glass. In 1896, Swedish scholar Svante Arrhenius concluded that fossil fuel combustion may eventually result in enhanced global warming. He predicted that a doubling of CO_2 concentration would lead to a 5 °C increase in global temperature. His warning was mostly ignored until the late 1950s. Roger R. D. Revelle and other pioneering scholars started scientific investigations on anthropogenic influences on global warming (*see* more detailed history in Archer and Pierrehumbert 2011). However, they hypothesized that oceans would absorb all the CO_2. Through the establishment of the International Geophysical Year (IGY) in 1958, the Atmospheric Carbon Dioxide Program was promoted to instrument two monitoring stations the Mauna Loa Observatory on Mauna Loa, Hawaii, and in Antarctica, with Charlie D. Keeling as the lead investigator. Later, the change in atmospheric CO_2 concentration became known as "the Keeling Curve" — a measure that is widely used in many current studies in global climate change and ecosystem science.

In 1967 Manabe and Wetherald (1967) published a groundbreaking paper based on the first generation of computer modeling. They stated that "The results show that it takes almost twice as long to reach the state of radiative convective equilibrium for the atmosphere with a given distribution of relative humidity than for the atmosphere with a given distribution of absolute humidity. Also, the surface equilibrium temperature of the former is almost twice as sensitive to change of various factors such as solar constant, CO_2 content, O_3 content, and cloudiness, as that of the latter, due to the adjustment of water vapor content to the temperature variation of the atmosphere." They concluded that "A doubling of the CO_2 content in the atmosphere has the effect of raising the temperature of the atmosphere (whose relative humidity is fixed) by about 2 °C." This early report set the stage for future investigations on the influences of CO_2 to global climatic change. In the late 1980s the scientific community realized the escalating increase in CO_2 concentration changed interdependently changes with global temperature. In 1988 the Intergovernmental Panel on Climate Change (IPCC) was founded by the United Nations Environmental Program (UNEP) and the

World Meteorological Organization (WMO) to assess the warming trend, explore the underline mechanisms, and predict future temperature under various scenarios. Quantifying the contributions of various greenhouse gases for their GWP has been the major task of the IPCC and the scientific community since then. For IPCC working groups, climate forcing is defined as "An externally imposed perturbation in the radiative energy budget of the Earth climate system, such as through changes in solar radiation, changes in the Earth albedo, or changes in atmospheric gases and aerosol particles." Climate forcing is the changes that can make current climate different.

In sum, greenhouse gases (GHGs) warm the Earth by absorbing energy and slowing the rate at which that energy escapes to outer space; they act like a blanket insulating the Earth. Different GHGs can have different effects on the Earth's warming (Fig. 5.2). Two key measures for assessing their effects are concentration, which determines the radiative efficiency of a GHG species, and lifespan (*i.e.*, how long it stays in the atmosphere). The contribution of each GHG species is quantified by calculating its GWP by converting its impact on CO_2 equivalence (*i.e.*, CO_2-equivalent) over a certain time horizon (TH) (*i.e.*, 20, 50, 100, 500 years, as applied in the IPCC reports). CO_2 has a lifespan of approximately 120 years, but it varies between 5 years and 200 years because its sources and sinks operate on many timescales while atmospheric sink is very small in the atmosphere; whereas CH_4 and N_2O have estimated lifespans of 12.4 and 120 years, respectively. In the literature, the term global warming impact (GWI) generally is used interchangeably with the more commonly used GWP.

5.1.3 The Roles of Terrestrial Ecosystems in GWP

Terrestrial ecosystems were hypothesized to play an important role in regulating the amount and dynamics of GHG (including water) as well as albedo. In the 1990s, after the First Assessment Report (FAR) of the IPCC (Houghton and Giddes 1992), a major hypothesis was that the "missing carbon" or unaccounted mismatch between accelerating fossil fuel emissions and the slower than expected growth rate of atmospheric CO_2 in the global carbon budget might be due to a significant underestimation of terrestrial carbon sink strength through photosynthesis (Sundquist 1993). Substantial research since 1990 has been conducted to quantify the carbon sequestration strength of global ecosystems. A parallel effort was made to quantify the amount of contributions from various GHGs (Lashof and Ahuja 1990, Rodhe

1990). Here, ecosystems are considered to have strong capability to add (N_2O, CH_4) or remove (CO_2, CH_4) GHGs from the atmosphere (Robertson *et al.* 2000, Syakila and Kroeze 2011, Knox *et al.* 2019). Chu *et al.* (2014) compared ecosystem fluxes of CO_2 and CH_4 in a soybean cropland and a nearby freshwater marsh in northwestern Ohio, USA. The average annual CH_4 flux was 49.7 g C-CH_4 $m^{-2}\,yr^{-1}$ from the marsh, which was compatible with its net annual CO_2 uptake (−21.0 g C-CO_2 $m^{-2}\,yr^{-1}$). But in the cropland the CH_4 flux was small and accounted for a minor portion of the atmospheric carbon budget. In coastal wetlands of Shanghai, Cheng *et al.* (2007) compared the CH_4 and N_2O fluxes of marshes dominated by native *Phragmites australis* and invasive *Spartina alterniflora.* They found that N_2O emission was higher in the invaded site than in native marshes. Additionally, CH_4 and N_2O emissions were suppressed and enhanced by clipping manipulations, respectively. Through a global synthesis of CH_4 fluxes from 60 monitoring tower sites, Knox *et al.* (2019) reported that annual estimates of net CH_4 flux ranged from −0.2 g C m^{-2} yr^{-1} for an upland forest site to 114.9 g C m^{-2} yr^{-1} for an estuarine freshwater marsh. The mean and median CH_4 fluxes were smaller at higher latitudes and larger at lower latitudes. At the LTER sites of the Kellogg Biological Station, Robertson *et al.* (2000) synthesized 10-year field measurements of CH_4 and N_2O in six experimental crops and found that none of the cropping systems provided net mitigation to GWP. When these measurements were integrated with CO_2 fluxes and different management and production options through life cycle analysis (LCA), the GHG impact was substantially higher than the emissions from an equivalent amount of fossil fuel-derived gasoline, including production, distribution, and combustion (Gelfand *et al.* 2011). Altogether, we know that all terrestrial ecosystems are N_2O sources and natural wetlands and rice paddies are also CH_4 sources, but upland soils are small CH_4 sinks (Neubauer and Megonigal 2015).

Another frontier in quantifying GWP of terrestrial ecosystems is resulted from the changes in albedo ($\Delta\alpha_s$) that reflect more solar radiation back to outer space (*e.g.*, lighter canopies) or keep more radiation energy within Earth systems (*e.g.*, taller and denser canopies) (Chapter 1). This is known as the radiative forcing (RF_s) of land surface. Imagine an area that can reflect 1% more solar radiation (*i.e.*, an increase of albedo by 1%). With an atmospheric transmission of 0.854, the total amount of solar energy that reaches the ground will be ~1170 W m^{-2} (*i.e.*, solar constant times 0.854). Assuming the same transmittance for the outgoing shortwave radiation, 1% more reflection is equivalent to ~10 W m^{-2} (*i.e.*, 1170×0.854×0.01). This value is equivalent to a cooling effect of ~11 kg CO_2, or ~3 kg C m^{-2} (*see*

Section 5.2 for conversion). Globally, the radiative forcing due to albedo changes has been most significant in regions with high snow cover or with land cover changes (conversion from forests to crop lands or urban lands). Aerosol contamination of snow can significantly reduce surface albedo, *a.k.a.* black-carbon, resulting in a positive feedback between warming-melting-albedo reductions (Hadley and Kirchstetter 2012). For example, strong negative trends in snow cover were observed in a study over North America and Eurasia (Déry and Brown 2007). Globally, Ghimire *et al.* (2014) reported an overall albedo increase of 0.00106 during 1700–2005, which is equivalent to -0.15 W m^{-2} cooling effect. IPCC (2013) claimed an overall cooling effect of 0.05–0.25 W m^{-2} using 289 ppmv as reference.

Land cover change (LCC) is another major process that has caused changes in land surface albedo. Across terrestrial landscapes albedo varies substantially by cover type and under different climate/weather conditions. In general, forests have a lower albedo than grasslands; water bodies have a much lower albedo than bare soils (*e.g.*, Chen *et al.* 2019). Cai *et al.* (2016) studied albedo effects in the context of expansive biofuel production in the United States. They found significant variations in albedo-induced effects among different land conversions, among crop systems, and among regions for the same land conversion. Yet, all conversions produced various degrees of cooling effects. In Europe, Carrer *et al.* (2018) reported that the introduction of cover crops into crop rotations during the fallow period would increase albedo by >4.17% of Europe's surface, which is equivalent to a mitigation of 15.91 g CO_2-eq yr^{-1} m^{-2}. Within a managed landscape, Sciusco *et al.* (2020) confirmed the differences in albedo among cover types and by climate, averaging between 0.4% and 2.0% in southwestern Michigan, USA. A landscape with a high proportion of forest can significantly reduce CO_2-eq mitigation by up to 24%–30%. Using our field measurements of albedo during the growing seasons as an example, a land conversion from forest to maize field will yield a cooling potential that is equivalent to -0.043 Mg C ha^{-1} due to a 0.051 increase in albedo, whereas a conversion to sorghum will result in a cooling equivalent to 0.094 Mg C ha^{-1} from an albedo increase of 0.111 (Fig. 5.3).

5.2 Calculating GWP of Greenhouse Gases

The global warming potential (GWP), also interchangeably called the global warming impact (GWI), is the most widely used metric for assessing the

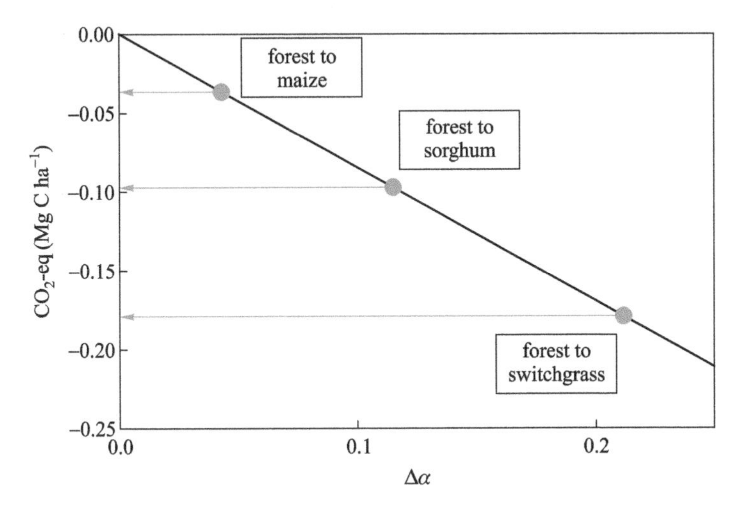

Fig. 5.3 Calculated CO_2-eq due to changes in albedo (α) that can result in either warming (positive values) or cooling (negative values) when a unit of land area (ha) is converted. Three examples of land conversions are based on our field measurements of albedo at the Kellogg Biological Station (*see* Section 5.3 for detailed calculations).

global warming impacts of different GHGs and albedo-induced changes. It was introduced by the IPCC in 1990 (Shine *et al.* 1990) to compile a national GHG inventory. It is designed to determine the lifetime and behavior of CO_2 or any other greenhouse gas in the atmosphere in varying time horizons (*i.e.*, 20, 50, 100 years) and allows landscape effects to be converted into CO_2 equivalences (CO_2-eq), or the emissions of CO_2 required to have the same climatic effect as the change being investigated. Based on the time integrated radiative forcing of a pulse emission of a unit mass of gas, GWP measures how much energy the emissions of 1 t of a gas will absorb over a given period of time, relative to the emissions of 1 t of CO_2 (Joos *et al.* 2013, Shine *et al.* 2005). The larger the GWP, the more a given GHG species warms the Earth. Its transparent algorithms, simplicity, and the relatively small number of input parameters required for the calculations make it highly favored by policymakers to compare emissions reduction, opportunities across sectors, and gases with little further input from scientists. However, it can be used incorrectly in some situations, especially in ecosystem management.

Although GWP has been lauded as a highly efficient method for weighing the climatic impact of emissions of different GHGs, it is not without some drawbacks. Despite its name, global warming potential does not represent the actual temperature change from each gas species but its time-integrated radiative forcing (Shine *et al.* 2005). This time integration (*i.e.*, the time horizon or TH) is very sensitive to short-lived GHGs (Levasseur *et al.* 2010). GWP does not effectively capture non-climatic effects such as ozone, aerosols, and carbon monoxide, due to their highly variable concentrations within the atmosphere. These forcing agents can have vastly distinct impacts on climate and the Earth system. While some GWP calculations can reflect preliminary changes in nitrogen due to yield and fertilizer, others do not account for methane from livestock production. Finally, GWP is also known to be affected by landscape dynamics such as natural disturbances (*e.g.*, wildfires, windthrows) and anthropogenic activities, such as land conversion (*e.g.*, forest to urban) and management practices (fertilization in agriculture, harvesting in forestry). Conversions of forested wetlands, for example, are of increasing concern to the scientific community because of large storage of carbon in ecosystems, high CO_2 emission through respiration, and high CH_4 emission when they are drained for other management objectives (*e.g.*, for forests or croplands). Depending on the climate and management, they can produce either cooling or warming impacts on the climate system. Frolking *et al.* (2006) modeled different climate and management in northern peatlands and concluded that the ratio of CH_4 emission to CO_2 sequestration was approximately 0.1–2.0 mol mol^{-1}, resulting a radiative forcing impact as a net warming that peaks after about 50 years for several hundred to several thousand years. Petrescu *et al.* (2015) synthesized the warming impacts from CO_2 and CH_4 when natural wetlands were converted to agricultural or forested lands. As expected, these conversions produced significant increases in atmospheric radiative forcing, particularly for wetlands converted to croplands. These two case examples highlight the complex nature of human influences, immediate feedbacks between ecosystems and the changing climate, as well as desynchronized consequences from different GHGs.

Carbon dioxide (CO_2), by definition, has a GWP value of 1 regardless of the time period used because it is the gas being used as the reference. Because CO_2 remains in the atmosphere for centuries, its emissions cause increases in atmospheric concentrations of CO_2 that can last thousands of years. Methane (CH_4), another common GWP gas, is estimated to have a GWP of 28–36 of CO_2 over 100 years. CH_4 emitted today has a much shorter time integration compared to CO_2 and N_2O, but it absorbs much more energy than CO_2. N_2O has an estimated GWP of 265–298 times that of CO_2 over a 100-year time horizon because it is much more potent in absorbing energy compared to the previous two gasses and has a long lifetime (Table 5.1). This is partly because CH_4 absorbs more radiation, but more importantly, its concentration in the atmosphere is low and not saturated. Note that impact of a gas grows logarithmically with concentration. CO_2

in the atmosphere is closer to saturation (*see* Fig. 5.4). Finally, chlorofluorocarbons (CFCs), hydrofluorocarbons (HFCs), hydrochlorofluorocarbons (HCFCs), perfluorocarbons (PFCs), and sulfur hexafluoride (SF_6) are sometimes called high-GWP gases because, for a given amount of mass, they absorb more longwave radiation than CO_2 and are extremely stable compounds that have thousands year lifetimes. Thankfully, most are in minute quantities in the atmosphere. However, the GWPs for these gases can be in the thousands or tens of thousands (Table 5.1).

Table 5.1 Global warming potential (GWP) of greenhouse gases (GHGs) based on the fourth Assessment Report (AR_4) of the IPCC (2007). The release of 1 kg CH_4, for example, is equivalent to 25 kg of CO_2 (from Forster *et al.* 2007).

GHG Species	Formula	Lifetime (years)	20-year	100-year	500-year
Carbon Dioxide	CO_2	Variable	1	1	1
Methane	CH_4	12±3	72	25	7.6
Nitrous Oxide	N_2O	120	289	298	153

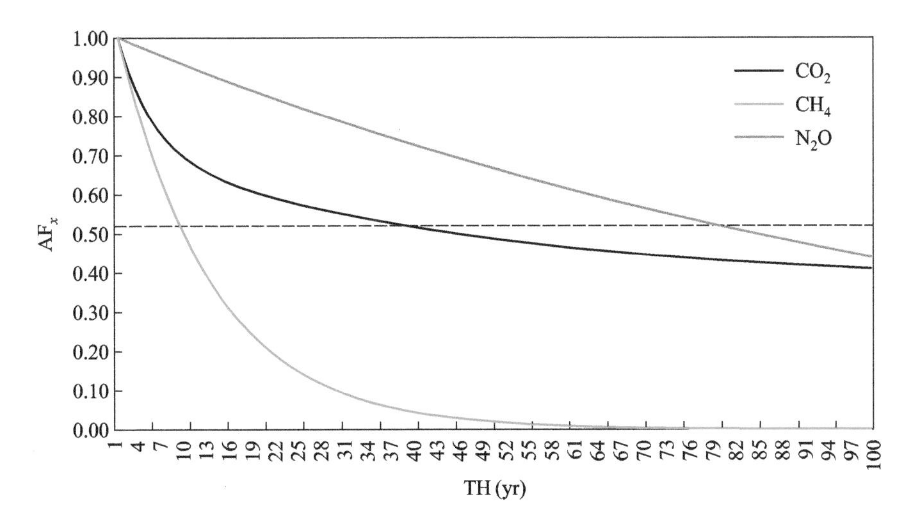

Fig. 5.4 The impulse response functions from the fifth IPCC Annual Report (AR_5) for calculating AF_x of CO_2, CH_4 and N_2O (Eqs. 5.5 and 5.6). The horizontal dashed line indicates the average AF_{CO_2} (0.52) over 100 years (TH = 100) (IPCC 2014 SM). This means that 52% of the emitted CO_2 pulse remains in the atmosphere after 100 years.

GWP is expressed with two units: CO_2-eq in mass (g, kg, or Mg) or radiative forcing (RF, W m^{-2}). This section describes the GWP induced by production of three major GHG species (CO_2, CH_4 and N_2O) and land use induced changes in albedo ($\Delta\alpha_s$) of an ecosystem. Over 20 parameters are needed to calculate GWP. Explanations and symbol uses for the major parameters are provided in Table 5.2. There are two approaches for calculating the global warming impacts of GHGs. The absolute global warming potential (AGWP) calculation takes into account both the infrared radiation absorption and the atmospheric lifetime (*e.g.*, atmospheric decay) of a gas. On the other hand, the relative GWP of a GHG is calculated as the ratio of its AGWP to that of CO_2. AGWP is calculated by integrating the radiative forcing due to GHG pulses over a time horizon (usually in kg W m^{-2} yr^{-1}). The AGWP of a GHG "x" pulse (GHG_x) is calculated as:

$$\mathrm{AGWP}_x(\mathrm{TH}) = \int_0^{\mathrm{TH}} \mathrm{IRF}_x(t) \cdot y(t)\mathrm{GHG}_x \tag{5.4}$$

where TH is the chosen time horizon (*i.e.*, 20, 100, 500 years), the impulse response function (IRF_x (t)) is the radiative forcing (W m^{-2}) of a gas "x" at time t, while $y(t)$ is proportion of the GHG "x" pulse that is still in the atmosphere at time t. This pulse is usually labelled AF (*i.e.*, Airborne Fraction). IRF_x represents the time-dependent radiative forcing caused by a specific GHG. This is determined by the following equation (IPCC 2013):

$$\mathrm{IRF}_x(t) = \exp\left(-\frac{t}{\tau_x}\right) \tag{5.5}$$

where IRF_x is the impulse response function of a specific GHG_x at the time horizon (t), t is the TH considered, and τ_x is the GHG_x lifetime in years (*e.g.*, CH_4 = 12.4 years and N_2O = 121 years).

The calculation of IRF for CO_2 is more complex because its lifetime cannot be represented by a simple exponential decay (Joos *et al.* 2013), partially because its lifespan is very difficult to define. Nevertheless, the following equation is proposed (IPCC 2013, Skytt *et al.* 2020):

$$\mathrm{IRF}_{\mathrm{CO_2}}(t) = a_0 + \sum_{i=1}^{3} a_i \cdot \exp\left(\frac{-t}{\tau_i}\right) \tag{5.6}$$

where $\mathrm{IRF}_{\mathrm{CO_2}}$ is the impulse response function of CO_2 at the time horizon (t), and t is the TH considered. The coefficients a_i and τ_i are constants and are equal to $a_0 = 0.2173$, $a_1 = 0.2240$, $a_2 = 0.2824$, $a_3 = 0.2763$, and $\tau_1 = 394.4$, $\tau_2 = 36.54$, $\tau_3 = 4.304$, respectively (Joos *et al.* 2013).

Equations (5.5) and (5.6) lead to the calculation of the fraction of GHG_x that remains in the atmosphere after the emission pulse, with losses occurring from either atmospheric chemical sinks, wet and dry deposition, and/or land

Table 5.2 Symbols, full names and their descriptions of parameters involved in calculating GWP of an ecosystem.

Symbol	Name	Description
CO_2	carbon dioxide	A colorless gas that occurs naturally in Earth's atmosphere
pCO_{2ref}	reference of partial CO_2 pressure	Reference of partial CO_2 pressure in the atmosphere, approximately at 389 ppmv or 0.383 g kg^{-1}
T_a	upwelling transmittance constant	Upwelling transmittance can be used as a constant based on a clear day (0.854), sunny day (0.73) or cloudy day (0.48), or it can be calculated based on zenith angle from time of day (Chapter 1)
TH	time horizon	The time horizon used when calculating GWP, the IPCC usually refers to TH 20 years, 100 years, and 500 years
α	albedo	The proportion of incident light to total solar radiation that is reflected by a surface. It is usually unitless but can be expressed as a percentage (*i.e.* 0.2 = 20%)
TOA	top of the atmosphere	The place where solar energy enters the Earth system, and where reflected light and thermal radiation from the sun-warmed Earth exit
S_w	shortwave	Shortwave radiation (UV/visible light) from solar energy within the atmosphere
RF_{CO_2}	radiative forcing of CO_2	Marginal RF of CO_2 emissions at the current atmospheric concentration. It is the derived radiative forcing from 1 kg of CO_2 at a constant of 0.908 (W kg^{-1} CO_2)
AF	airborne fraction	Proportion of human-emitted GHG that remains in the atmosphere after a certain period of time. AF of CO_2 is 0.69, 0.48 and 0.32 for time horizons of 20, 100, and 500 years, respectively. These values refer to other calculations (Bright *et al.* 2015)
S	local area	The area affected by the change in surface albedo (m^{-2})
$\Delta\alpha_s$	delta albedo	The local change in albedo in a region at a specific time. Calculated by subtracting a reference albedo
$RF_{\Delta\alpha}$	albedo-induced radiative forcing	The change in net radiative flux from the surface driven by surface albedo changes ($\Delta\alpha$)
ΔRF_{TOA}	delta radiative forcing at the top of the atmosphere	The RF at the TOA due to surface albedo changes. It is usually a function of latitude and it is related to changes in α

continued

Symbol	Name	Description
$GWP_{\Delta\alpha}$	global warming potential	Also called global warming impact (GWI), it is a quantified measure of the relative radiative forcing impact due to albedo change, specific to CO_2
M_{CO_2}	molecular weight of CO_2	The atomic weight of carbon is 12, while oxygen is 16. Therefore, the molecular weight of CO_2 is 44 g mol^{-1} (Table 5.4)
m_{air}	atmospheric mass	The total mean mass of the atmosphere, which is 5.1480×10^{18} kg
S_{Earth}	area of Earth	The surface area of the Earth, which is ~510 million km^2 (5.1×10^8 km^2)
ΔF_{2x}	forcing of CO_2 concentration	Radiative forcing resulting from a doubling of current CO_2 concentration in the atmosphere, at a current rate of +3.7 W m^{-2}
M_{air}	molecular weight of dry air	The weight of dry air (28.95 g mol^{-1}). It is composed of nitrogen (78%), oxygen (20.85%), argon (0.93%) and other gases (0.04%)

and ocean absorption. This is often called the airborne fraction (AF_x) of a GHG_x. It is expressed in percentage or (0–1) (Fig. 5.4). AF_x tends to decrease over the years. The literature often uses a constant AF_x value, such as the average AF_x for a specific TH (Betts 2000, Akbari *et al.* 2009). For example, the AF_{CO_2} at TH = 100 is normally set at 0.52 (or 52%).

Most GHGs are involved in extremely complex chemical interactions within the atmosphere. Thus, assuming that the GHG (*e.g.*, CO_2) is distributed evenly in the atmosphere, the IRF of a model is then normally computed by monitoring the decrease of an initial atmospheric CO_2 perturbation due to a pulse-like CO_2 release into an atmosphere. This approach is useful, as it represents GHGs in climate models as they change temporally, which is shown in the following equation:

$$IRF_{CO_2} = \frac{RF \cdot M_C}{C_0 \cdot GTC \cdot M_{CO_2}} \tag{5.7}$$

where RF is the radiative forcing constant for CO_2 (5.35 W m^{-2}), C_0 is the reference concentration (389 ppm), GTC is the conversion from 1 ppm of atmospheric CO_2 to equivalent gigaton carbon (2.123E+12 kg C ppm^{-1}), M_C is the molecular weight of carbon (12 g mol^{-1}), and M_{CO_2} is themolecular weight of CO_2 (44 g mol^{-1}). Multiplication of the IRF_{CO_2} and AF yields a value of 1.767E−15 kg W m^{-2} $CO_2{}^{-1}$. The RF value represents the sensitivity of longwave radiation to CO_2 concentration (*i.e.*, feedbacks to warming potential); its values are conventionally estimated from climate models.

To calculate AGWP of CH_4 and N_2O we use a modified version of Equation (5.4) (IPCC 2013). For further information, *see* Myhre *et al.* (2013) and Skytt *et al.* (2020).

The IPCC currently reports the AGWP for common GHGs at 20-year and 100-year intervals. The $AGWP_{100}$ for CO_2 is 9.351E−14, while CH_4 is 2.61E−12 and N_2O is 2.43E−11. These values enable a smoother calculation of GWP by dividing the $AGWP_x$ by the AGWP of a reference gas. To calculate the relative GWP of a gas "x", the following equation is then used:

$$\mathrm{GWP}_x(\mathrm{TH}) = \frac{\mathrm{AGWP}_x(\mathrm{TH})}{\mathrm{AGWP}_{\mathrm{CO_2}}(\mathrm{TH})} \tag{5.8}$$

In the case of CO_2, GWP can also be derived by determining the RF_{CO_2}, which uses RF_{CO_2} to analyze the fraction of CO_2 remaining in the atmosphere after a single pulse emission from interactions between the atmosphere, oceans, and the terrestrial biosphere (Joos *et al.* 2013). It is determined by calculating the time-integrated atmospheric response function of CO_2 with its radiative forcing and converting it into equivalent CO_2 using the following equation:

$$\mathrm{RF}_{\mathrm{CO_2}} = \frac{\ln(2) \cdot p\mathrm{CO}_{2\mathrm{ref}} \cdot M_{\mathrm{CO_2}} \cdot m_{\mathrm{air}}}{S_{\mathrm{Earth}} \cdot \Delta F_{2x} \cdot M_{\mathrm{air}}} \tag{5.9}$$

where $p\mathrm{CO}_{2\mathrm{ref}}$ is a reference of partial CO_2 pressure in the atmosphere (389 ppmv or 0.383 kg g^{-1}), S_{Earth} is the area of the Earth's surface (5.1×10^{14} m^2), $M_{\mathrm{CO_2}}$ is the molecular weight of CO_2 (44.01 g mol^{-1}), m_{air} is the mass of the atmosphere (5.148×10^{18} kg), ΔF_{2x} is the radiative forcing resulting from a doubling of current CO_2 concentration in the atmosphere (+3.7 W m^{-2}), and M_{air} is the molecular weight of dry air (28.95 g mol^{-1}). The inverse of this equation then provides us with a constant of 0.908 (W kg ${CO_2}^{-1}$) (Table 5.1), which can then be compared to other sources of CO_2 emissions such as CO_2 fluxes in agriculture (Cherubini *et al.* 2011, Gelfand *et al.* 2011). AF is the ratio of the annual increase in atmospheric CO_2 to the total CO_2 emissions, and its value has been disputed by many researchers — variously estimated at 0.53 (Joos *et al.* 2013), 0.55 (Akbari 2009) and 0.52 (Bright *et al.* 2015). However, most agree on a 20% uncertainty in predicting the amount of CO_2 remaining in the atmosphere due to variations in Earth's land and ocean carbon sinks from changing weather and emission patterns (Table 5.3).

Calculated changes in RF value (*i.e.*, ΔRF) are used in the calculation of AGWP (IPCC 2013). For each GHG, past and present global abundance (in

Table 5.3 Radiative forcing (RF) calculations used for calculating Annual Greenhouse Gas Index (AGGI) by NOAA[1]. These empirical expressions are derived from atmospheric radiative transfer models and generally have an uncertainty of ~10%. The uncertainties in the global average abundances of the long-lived greenhouse gases are much smaller (<1%).

GHG	ΔF (W m^{-2})	Constant	RF (W m^{-2})
CO_2	$\Delta F = \alpha \ln (C/C_o)$	$\alpha = 5.35$	5.35
CH_4	$\Delta F = \beta(M^{1/2} - M_o^{1/2}) - [f(M, N_o) - f(M_o, N_o)]$	$\beta = 0.036$	2.7
N_2O	$\Delta F = \varepsilon(N^{1/2} - N_o^{1/2}) - [f(M_o, N) - f(M_o, N_o)]$	$\varepsilon = 0.12$	0.65

*IPCC (2001); The subscript $_o$ denotes the unperturbed (1750) global abundance; $f(M, N) = 0.47\ln[1 + 2.01\times10$ $-5\ (MN)0.75 + 5.31\times10\text{-}15M(MN)1.52]$; C is the CO_2 global measured abundance in ppm, M is the same for CH_4 (ppb); N is the same for N_2O in ppb, X is the same for CFCs (ppb); $C_o = 278$ ppm, M_o=722 ppb, N_o=270 ppb, X_o=0.

ppm for CO_2, in ppb for CH_4 and N_2O) as well as constant values are considered to calculate the radiative forcing of the GHG_x (Table 5.3). Regarding CH_4, augmented radiative forcing can be calculated by multiplying the constant β by the value 1.65, which includes the indirect effects of methane on the ozone and stratospheric H_2O:

$$\mathrm{GWP} = \frac{\Delta \mathrm{RF_{TOA}} \cdot S}{\mathrm{AF} \cdot \mathrm{RF_{CO_2}}} \cdot \frac{1}{\mathrm{TH}} \tag{5.10}$$

where, ΔRF_{TOA} is the radiative forcing (W m^{-2}) from changes in albedo from the TOA (Eq. 5.9), S is the local area subjected to albedo change (m^{-2}), AF is percentage of emitted CO_2 that remains in the atmosphere after a period of time from anthropogenic sources, RF_{CO_2} is the derived radiative forcing from 1 kg of CO_2, and TH is the time horizon for 100 years (TH = 100). See Table 5.2 for detailed explanations of abovementioned parameters and Table 5.4 for user defined values in calculating GWP.

It must be stressed that there is no universally accepted methodology for combining all required components into a single GWP. Multiple angles have been approached in the last decade for determining the most efficient method for relating and modelling all necessary variables for the sake of simplicity. For example, Bright *et al.* (2013, 2015) and Sieber *et al.* (2019) used AGWP that is determined by integrating radiative forces over a specific time horizon (*i.e.*, 100 years) (Table 5.1), while Muñoz *et al.* (2010) and Carrer *et al.* (2018) attempted to determine an impulse response function (IRF) using a slightly different method to achieve a constant paired with an airborne

1 https://www.esrl.noaa.gov/gmd/aggi/aggi.html

Table 5.4 User defined variables for the calculation of GWP.

Variable	Value	Unit
M_{CO_2}	44.01	g mol^{-1}
m_{air}	5.15E+18	kg
S_{Earth}	5.10E+14	m^2
ΔF_{2x}	3.7	W m^{-2}
M_{air}	28.95	g mol^{-1}
M_{CO_2}	44.01	g mol^{-1}
AF	0.53	unitless
S	1	m^2
$\Delta\alpha_s$		unitless
RF_s		W m^{-2}
ΔRF_{TOA}		W m^{-2}
$GWI_{\Delta\bar{\alpha}_s}$		kg CO_2 yr^{-1}
TH	100	year
T_a	0.854	unitless
C/CO_2	0.273	
N	1	day
RF_{CO_2}	0.908	W kg CO_2^{-1}
$\Delta F_{2x}(CO_2)$	3.7	W m^{-2}

fraction, which inherently achieves the same result.

5.3 Calculating GWP from Surface Albedo

To establish direct comparisons between changes in surface albedo ($\Delta\alpha_s$) and its effects on a landscape to GWP, radiative forcing or albedo can be converted into CO_2-eq. To first determine α_s, the ratio of reflected light ($Sw_\uparrow$) to the total incident sunlight ($Sw_\downarrow$) for a given area of land surface is calculated as albedo:

$$\alpha_s = \frac{Sw_\uparrow}{Sw_\downarrow} \tag{5.11}$$

Albedo is angle dependent and wavelength dependent. Black sky albedo (*i.e.*, direct light albedo) differs from white sky albedo (*i.e.*, diffuse albedo). Albedo under real-world conditions with a combination of both black and white sky albedo is called blue sky albedo. The global surface albedo is the ratio of global incoming shortwave radiation divided by outgoing, averaged over all angles, wavelengths, and cloud illumination conditions. Across the terrestrial lands, albedo varies from 0 to 1. Fresh snow or a mirror have albedo of 0.8–1.0 while water or asphalt have albedos near 0.1. On average, global annual albedo is approximately 0.3 (Goode *et al.* 2001), accounting for land, ocean, and clouds. Because albedo is only derived during sunlight hours, it is imperative to ensure radiation values are measured during these hours (*i.e.*, daytime, Chapter 1). Contribution of albedo to GWP is also meaningful only when compared with a reference to the region (*e.g.*, a site without human disturbances, or historical values). Change in albedo ($\Delta\alpha_{\mathrm{s}}$) can then be found by determining the difference between the albedo value of the reference site and the target ecosystem:

$$\Delta\alpha_{\mathrm{s}} = \alpha_{\mathrm{sNEW}} - \alpha_{\mathrm{sREF}} \tag{5.12}$$

During the summer, $\Delta\alpha_{\mathrm{s}}$ are much more pronounced among ecosystem types due to seasonal high solar irradiance and zenith angle and vegetation growth. In agricultural regions, $\Delta\alpha_{\mathrm{s}}$ is highly dependent of surface moisture, crop height, planting density, crop species, crop cover and land management practices (*e.g.*, irrigation, fertilization, harvest, tilling, *etc.*). However, in the winter, $\Delta\alpha_{\mathrm{s}}$ in temperate zones is usually not significant because landscapes are covered by snow or are bare. Once $\Delta\alpha_{\mathrm{s}}$ is found, it can then be used to calculate the instantaneous radiation forcing ($\mathrm{RF}_{\Delta\alpha}$) of:

$$\mathrm{RF}_{\Delta\alpha} = -Sw_{\downarrow}\Delta\alpha_{\mathrm{s}} \tag{5.13}$$

Finally, $\mathrm{RF}_{\Delta\alpha}$ can be inserted into the GWP equation via IRF method (Eq. 5.14) and AGWP method (Eq. 5.15):

$$\Delta\mathrm{RF}_{\mathrm{TOA}} = -\frac{1}{N}T_{\mathrm{a}} \cdot \mathrm{RF}_{\Delta\alpha} \tag{5.14}$$

$$\Delta\mathrm{RF}_{\mathrm{TOA}} = -\frac{1}{N}\frac{S \cdot T_{\mathrm{a}} \cdot \mathrm{RF}_{\Delta\alpha}}{S_{\mathrm{Earth}}} \tag{5.15}$$

where $\mathrm{RF}_{\Delta\alpha}$ is the change in net radiative forcing from the surface driven by surface albedo (W m^{-2}), $Sw_{\downarrow}$ is local incoming solar radiation incident to the surface (W m^{-2}), T_{a} is the upwelling transmittance derived from estimating thermal radiant fluxes within the environment (Campbell and Norman 2012; Chapter 1), N is the number of days (or hours), and $\Delta\alpha_{\mathrm{s}}$ is the local change in albedo between two specific surfaces. The value of the ΔRF_{TOA} is a representation of the daily local power (W m^{-2}) that would have been reflected back to outer space due to changed land surface (*e.g.*, vegetation cover, greenness, moisture; Carrer *et al.* 2018). Positive and negative RFs due to $\Delta\alpha_s$ correspond to warming and cooling effects, respectively, which can determine whether an ecosystem is actively mitigating global warming (Caiazzo 2014). Regardless of which method (AGWP or GWP) is used, surface albedo can be determined over a specific area to determine GWP for a spatiotemporal period.

5.4 Case Examples

Once the model parameters are determined, calculations of GWP are straightforward. Here we provide two examples as demonstrations.

Example I. Find the GWP for CO_2, CH_4, and N_2O for a switchgrass (*Panicum virgatum*) site with $Sw_\downarrow$ of 478 W m^{-2} and $Sw_\uparrow$ of 134 W m^{-2}, when the reference native grasses have a $Sw_\downarrow$ of 482 W m^{-2} and a $Sw_\uparrow$ of 83 W m^{-2}. The T_a of the day is 0.854. $AGWP_{100}$ for CO_2 is 9.351E−14, while CH_4 is 2.61E−12 and N_2O is 2.43E−11.

Solution using AGWP method:

Switchgrass $\alpha_s : 134/478 = 0.28$
Native grasses $\alpha_s : 83/482 = 0.17$
$\Delta\alpha_s = 0.28 - 0.17 = 0.11$
$RF_{\Delta\alpha} = -(478 \times 0.11) = -52.6 \text{ W m}^{-2}$
$\Delta RF_{TOA} = -\{1/1 \times [1 \times 0.854 \times (-52.6)/(5.10E-14)]\} = -8.80E-14 \text{ W m}^{-2}$
$GWP = -(-8.80E-14)/(9.17E-14) = -9.39E-01 \text{ kg } CO_2 \text{ m}^{-2} \text{ yr}^{-1}$

Solution using IRF method:

Switchgrass $\alpha_s : 134/478 = 0.28$
Native grasses $\alpha_s : 83/482 = 0.17$
$\Delta\alpha_s = 0.28 - 0.17 = 0.11$
$RF_{\Delta\alpha} = -(478 \times 0.11) = -52.6 \text{ W m}^{-2}$
$\Delta RF_{TOA} = -[1/1 \times (0.854 - 52.6)] = -44.9 \text{ W m}^{-2}$
$GWP = [(-44.9 \times 1)/(0.52 \times 0.908) \times (1/100)]$
$= -9.39E-01 \text{ kg } CO_2 \text{ m}^{-2} \text{ yr}^{-1}$

GWP *for each GHG*:

$GWP_{CO_2} = (-8.80E-14)/(9.17E-14) = -9.39E-01 \text{ kg } CO_2 \text{ m}^{-2} \text{ yr}^{-1}$

$$\mathrm{GWP_{CH_4}} = (-8.80\mathrm{E}-14)/(2.61\mathrm{E}-12) = -3.31\mathrm{E}-02 \text{ kg } \mathrm{CH_4 \ m^{-2} \ yr^{-1}}$$
$$\mathrm{GWP_{N_2O}} = (-8.80\mathrm{E}-14)/(2.43\mathrm{E}-11) = -3.55\mathrm{E}-03 \text{ kg } \mathrm{N_2O \ m^{-2} \ yr^{-1}}$$

The solutions above not only prove both methods of calculating GWP, which have been utilized in the literature over the past decade, but also show how to integrate other GHG emissions in context of GWP. As mentioned in Section 5.2, the reference of GWP for CO_2 is 1, while the GWP of CH_4 and N_2O is approximately 28 and 264, respectively (Table 5.1). These magnitudes are compared in Table 5.5.

Table 5.5 Explanation of GWP in relation to reference GWP_{CO_2}.

GHG Species	Solution GWP (kg $CO_2/CH_4/N_2O$ m^{-2} yr^{-1}	GWP Reference	GWP Reference Magnitude to CO_2
CO_2	−9.39E−01	1	1.00
CH_4	−3.31E−02	28	28.41
N_2O	−3.55E−03	264	264.50

One of the critical parameters in calculating GWP involves the bias from the upwelling transmittance constant (T_a). T_a is usually derived from clear-sky conditions, often as a constant of 0.854 (Lenton and Vaughan 2009, Bright and Kvalevåg 2015). This value represents regions with very little cloud cover, such as deserts (Muñoz *et al.* 2010). However, a T_a value of 0.854 usually results in an overestimation of $RF_{\Delta\alpha}$(Carrer *et al.* 2018) in regions with highly variable weather or persistent cloud cover. Liu and Jordan (1960) determined that T_a can be calculated for other typical sky conditions, such as overcast ($T_a < 0.45$), sunny ($T_a < 0.75$) and clear ($T_a > 0.85$). The upwelling transmittance can also be determined on a daily basis, using the solar zenith angle and latitude (S1-5: Solar.PY in Chapter 1) and ground measurements, which will reduce the amount of error and bias in the calculations. This method is known to reduce bias up to at least 30% of $RF_{\Delta\alpha}$ (Sciusco *et al.* 2020) in northern forested and grassland regions. To calculate T_a, the following equation is needed:

$$T_a = \frac{Sw_{\downarrow}}{\mathrm{SW_{TOA}}} \tag{5.16}$$

where SW_{TOA} (W m^{-2}) is the value of extraterrestrial radiation energy falling on a canopy surface. To determine SW_{TOA}, three constants should be found based on the day of the year (DOY):

$$\mathrm{SW_{TOA}} = I_{sc} \cdot d_r \cdot I_\theta \tag{5.17}$$

where I_{sc} is the solar constant (1367 W m^{-2}), I_θ is the extraterrestrial irradiance intensity onto a crop canopy $[\cos(\theta_{ZENITH} \cdot \frac{\pi}{180})]$, and d_r is the Earth-sun distance $[1+0.0334\cos(\text{DOY} \cdot \frac{2\pi}{365.25})]$. Alternative equations are also provided in Chapter 1.

Example II. Calculations of an upwelling transmittance value (T_a) for the 27th July, 2018, with a $Sw_\downarrow$ value of 691.55 W m^{-2} are listed below:

$$\text{SW}_{\text{TOA}} = 1367 \text{ W m}^{-2} \cdot \left[1 + 0.0334\cos\left(\text{DOY}\frac{2\pi}{365.25}\right)\right] \cdot \left(\cos 41.89 \cdot \frac{\pi}{180}\right)$$

$$= 980.99 \text{ W m}^{-2}$$

$$T_a = \frac{691.55}{980.99} = 0.70$$

A spreadsheet model for calculating AGWP and GWP is provided as a supplementary document (S5-2: GWP_model.xlsx) with this chapter. Application of this model needs actual fluxes of CO_2, CH_4 and N_2O in μmol m^{-2} s^{-1}, as well as the total incoming radiation (S). It also needs the albedo value of the study ecosystem and a reference site for calculating the difference (*i.e.*, $\Delta\alpha$) for albedo-induced GWP. The GWP in several unit expressions are provided, including radiation forcing in W kg^{-1} CO_2, GWP of CO_2-eq in kg m^{-2} yr^{-1}, C-eq in Mg C ha^{-1} yr^{-1}, and biomass-equivalent in Mg ha^{-1} yr^{-1}.

To further demonstrate the model application, we used the ground measurements of CO_2, N_2O and CH_4 fluxes and albedo of four bioenergy crops at the Kellogg Biological Station (KBS), Michigan, USA, for GWP calculations. The reference site is a mixed prairie under the Conservation Reserve Program (CRP). Three crop ecosystems are a continuous corn plantation, switchgrass, and managed prairies (Zenone *et al.* 2011). These sites were instrumented with eddy-covariance flux towers for CO_2 flux, albedo minoring sensors, and static chambers for CH_4 and N_2O emission/uptake. The mean values of the three GHG species and albedo are based on Robertson *et al.* (2000) and Abraha *et al.* (2019) for calculating the GWPs of the three ecosystems (Table 5.6).

Calculated GWPs for these sites are similar to those of Abraha *et al.* (2018, 2019) and Gelfand *et al.* (2011). In brief, the switchgrass crop offsets the highest GWP (6.14 Mg C ha^{-1} yr^{-1}) because of its high CO_2 sequestration and high albedo. Here the GWP of the high albedo is 4.18 (Mg C ha^{-1} yr^{-1}), which is 20.1% of ecosystem production of carbon. This value alone is much higher than the total GWP of other three crop types. The net GWPs of the corn and the mixed prairie indicate they provide warming and cooling services to the atmosphere, respectively, though the magnitude (*i.e.*, 0.028

Table 5.6 Annual mean fluxes of CO_2, N_2O and CH_4 as well as from the albedo differences of three bioenergy crops at the KBS for calculating GWP (Eqs. 5.4–5.15). The input parameters were generated based on the literature of Abraha *et al.* (2019) and Robertson *et al.* (2000) in assessing the contributions of each GHG toward ecosystem GWP. Albedo values were from our *in situ* measurements in 2018.

GHG flux and GWP	Reference	Corn	Switchgrass	Mixed Prairies
Mean CO_2 (μmol m^{-2} s^{-1})	1.823	0.4920	5.485	0.0789
N_2O (μmol m^{-2} s^{-1})	0.000018	0.0025	0.00045	−0.000022
CH_4 (μmol m^{-2} s^{-1})	0.00064	0.00078	0.00127	0.00663
Albedo (α)	0.12	0.17	0.33	0.18
$\Delta\alpha$	0.00	0.05	0.21	0.06
GWP-CO_2 (Mg ha^{-1} yr^{-1})	−6.907	−1.864	−20.782	−0.299
GWP-CH_4 (Mg ha^{-1} yr^{-1})	−0.025	−0.031	−0.050	−0.260
GWP-N_2O (Mg ha^{-1} yr^{-1})	0.018	2.506	0.451	0.022
GWP-$\Delta\alpha$ (Mg ha^{-1} yr^{-1})	0.000	−0.498	−4.180	−0.597
GWP -Total (Mg ha^{-1} yr^{-1})	−1.729	0.028	−6.140	−0.284

and 0.284 Mg C ha^{-1} yr^{-1}) is at a similar or much higher level compared to the global mean net ecosystem production of grasslands (*i.e.*, ~ 0.1 Mg C ha^{-1} yr^{-1}) or to the reference site 1.73 (Mg C ha^{-1} yr^{-1}).

Finally, reforestations of marginal lands/croplands are become the most used global effort to improve ecosystem functions, including GWP reduction. Here we provide another spreadsheet model to demonstrate the changes in GWP over a 100-year period (Fig. 5.5). An exponential curve is used to simulate the changes in albedo and the net ecosystem exchanges of CO_2, CH_4 and N_2O based on estimated values from the literature (*e.g.*, Abraha *et al.* 2018, Robertson *et al.* 2000). The default parameters for CO_2 flux are set with an initial net carbon loss of −1.5 Mg C ha^{-1} yr^{-1} and a maximum gain of 2.0 Mg C ha^{-1} yr^{-1}. These values are approximated with average net ecosystem production (NEP) of deciduous forests in northern America (Amiro *et al.* 2010). The shape factor was set at 0.05 by assuming that the forest would reach its maximum NEP in 30–40 years. For CH_4 and N_2O fluxes, we generated the average flux from Robertson *et al.* (2000) for the forest and a no-till cropland. The shape factors were set at 0.15 and 0.12, respectively, allowing the fluxes reaching level of forest at about 15–20 years. Initial albedo values were based on our field measurements, and they were set at 0.131 for the cropland and 0.102 for the forest; this way, the cropland

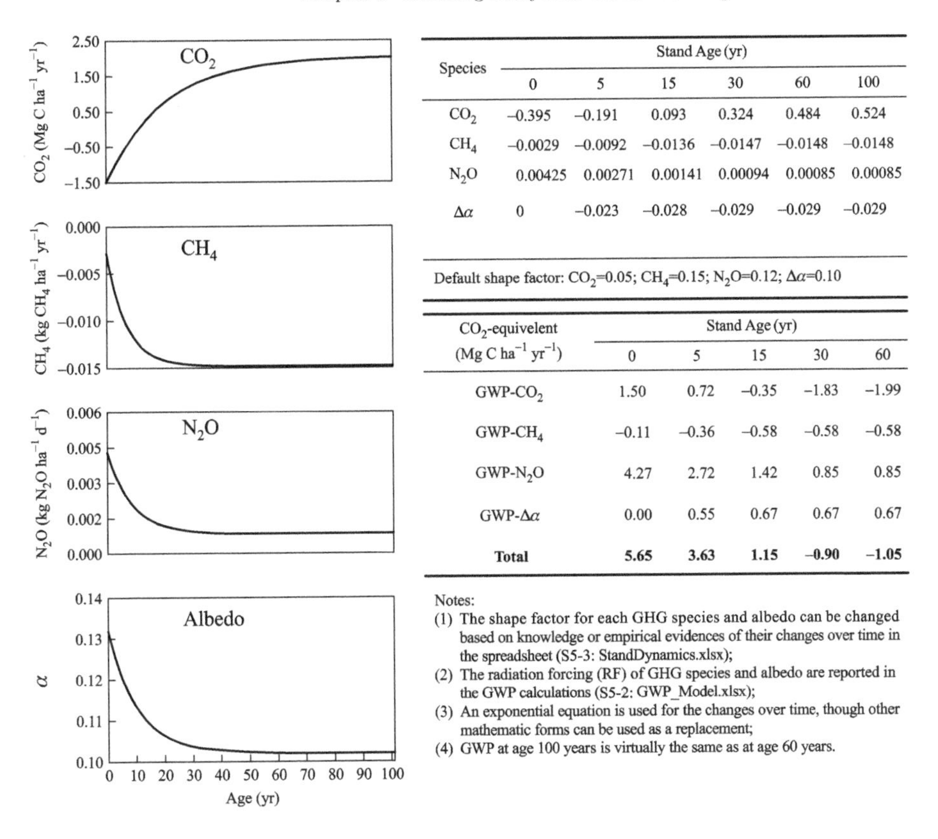

Species	Stand Age (yr)					
	0	5	15	30	60	100
CO_2	−0.395	−0.191	0.093	0.324	0.484	0.524
CH_4	−0.0029	−0.0092	−0.0136	−0.0147	−0.0148	−0.0148
N_2O	0.00425	0.00271	0.00141	0.00094	0.00085	0.00085
$\Delta\alpha$	0	−0.023	−0.028	−0.029	−0.029	−0.029

Default shape factor: CO_2=0.05; CH_4=0.15; N_2O=0.12; $\Delta\alpha$=0.10

CO_2-equivelent (Mg C ha^{-1} yr^{-1})	Stand Age (yr)				
	0	5	15	30	60
GWP-CO_2	1.50	0.72	−0.35	−1.83	−1.99
GWP-CH_4	−0.11	−0.36	−0.58	−0.58	−0.58
GWP-N_2O	4.27	2.72	1.42	0.85	0.85
GWP-$\Delta\alpha$	0.00	0.55	0.67	0.67	0.67
Total	**5.65**	**3.63**	**1.15**	**−0.90**	**−1.05**

Notes:
(1) The shape factor for each GHG species and albedo can be changed based on knowledge or empirical evidences of their changes over time in the spreadsheet (S5-3: StandDynamics.xlsx);
(2) The radiation forcing (RF) of GHG species and albedo are reported in the GWP calculations (S5-2: GWP_Model.xlsx);
(3) An exponential equation is used for the changes over time, though other mathematic forms can be used as a replacement;
(4) GWP at age 100 years is virtually the same as at age 60 years.

Fig. 5.5 Simulated changes in albedo and fluxes of three GHGs over a 100-year period. The model parameters used for calculating GWP of each species are from the literature (*top table*). GWP from each GHG and albedo are simulated using the S5-2 (GWP_Model.xlsx) for six ages during the course of 100 years (*bottom table*).

level reaches the forest level at 20–30 years. Both the initial/ending values and the rate of change can be altered by users for specified applications. The exponential model for predicting the changes can also be replaced with other equation forms.

The CO_2-equivelent values due to changes in albedo and the three GHG species were predicted for stand age of 0, 5, 15, 30, 60 and 100 years as examples by using the GWP spreadsheet model (S5-2: GWP_Model.xlsx). The forest would be net carbon source, measured by total CO_2-equivelent of GWP, until age 30. During this period, the GWP of CO_2-equivelent is attributable to high CO_2 and N_2O emissions, with $\Delta\alpha$ and CH_4 setting off the GWP. Over the time, CO_2 uptake becomes the dominant process turning the forest into a net carbon sink regardless of lowered albedo that elevates GWP values (up to 0.67 Mg C ha^{-1} yr^{-1}). A small reduction in

N_2O in later stages also contributes to a significant reduction in total GWP (from 4.27 Mg C ha^{-1} yr^{-1} to 0.85 Mg C ha^{-1} yr^{-1}). It is notable that the contributions of albedo changes to the total GWP appear to be similar to the level of CH_4-induced GWP, due to its high radiative forcing. An alternative interpretation of these results is to compare forests of different ages across a landscape, assuming they have similar land use history. For example, a 5-year-old stand would have a CO_2-equivalent of 3.63 Mg C ha^{-1} yr^{-1}, compared to -1.05 Mg C ha^{-1} yr^{-1} for a 60-year-old mature forest.

5.5 Summary

Investigating global warming potential (GWP) of various ecosystems, especially those under human influences, is becoming a core research focus in ecological studies. After brief reviews of the physical foundation of energy balance for the Earth system, we provided a short history of global warming and anthropogenic influences. Then we followed the IPCC protocols and provided step-by-step calculations of actual and relative GWP. For a better understanding the algorithms involved in the calculations, physical and chemical background of energy balance, as well as the role of greenhouse gases, are provided as an introduction. A spreadsheet model for calculating GWP of an ecosystem is provided for practical uses. Examples from experimental sites at the Kellogg Biological Station are used for demonstrations.

Online Supplementary Materials

S5-1: Solar radiation intensity and absorptions by different gases through the atmosphere. Data provided by Robert Rhode of Berkeley Earth (AtmosphereTransmission.txt).
S5-2: Spreadsheet model for calculating actual global warming potential (AGWP) and the relative GWP based on *in-situ* input parameters (GWP_Model.xlsx).
S5-3: Simulations of GWP changes over a 100-year period when a stand is planted for forest from managed cropland (StandDynamics.xlsx).

Scan the QR code or go to
https://msupress.org/supplement/BiophysicalModels
to access the supplementary materials.
User name: biophysical
Password: z4Y@sG3T

Acknowledgements

We appreciate the long-term support, encouragements and stimulating discussion from Phil Robertson in modeling global warming potentials. Michael Abraha provided timely support for accessing the data at the GLBRC sites of the Kellogg Biological Station. Robert Rhode of Berkeley Earth provided the original image and re-analyzed data for spectral distribution of solar radiation intensity and absorptions. Pengsheng Wei explained the data source of GHG absorptions and connected us with Rhode's original work. Gordon Smith, Ankur Desai and Phil Robertson provided constructive suggestions for improving the draft manuscript. Kristine Blakeslee edited the language and format of the draft.

References

Abraha, M., Gelfand, I., Hamilton, S. K., Chen, J., and Robertson, G. P. (2018). Legacy effects of land use on soil nitrous oxide emissions in annual crop and perennial grassland ecosystems. *Ecological Applications*, 28(5), 1362–1369.

Abraha, M., Gelfand, I., Hamilton, S. K., Chen, J., and Robertson, G. P. (2019). Carbon debt of field-scale conservation reserve program grasslands converted to annual and perennial bioenergy crops. *Environmental Research Letters*, 14: 024019.

Akbari, H., Menon, S., and Rosenfeld, A. (2009). Global cooling: Increasing world-wide urban albedos to offset CO_2. *Climatic Change*, 94(3–4), 275–286.

Amiro, B. D., Barr, A. G., Barr, J. G., Black, T. A., Bracho, R., Brown, M., Chen, J., Clark, K. L., Davis, K. J., Desai, A. R., Dore, S., Engel, V., Fuentes, J. D., Goulden, M. L., Kolb, T. E., Lavigne, M. B., Law, B. E., Margolis, H. A., Martin, T., McCaughey, J. H., Montes-Helu, M., Noormets, A., Randerson, J. T., Starr, G., and Xiao, J. (2010). Ecosystem carbon dioxide fluxes after disturbance in forests of North America. *Journal of Geophysical Research: Biogeosciences*, 115(G4), Gooko2.

Archer, D., and Pierrehumbert, R. (2011). *The Warming Papers: The Scientific Foundation for the Climate Change Forecast.* John Wiley and Sons, 419 pp.

Betts, R. A. (2000). Offset of the potential carbon sink from boreal forestation by decreases in surface albedo. *Nature*, 408(6809), 187–190.

Bright, R. M., and Kvalevåg, M. M. (2013). Technical note: Evaluating a simple parameterization of radiative shortwave forcing from surface albedo change. *Atmospheric Chemistry and Physics*, 13(22), 11169–11174.

Bright, R. M., Zhao, K., Jackson, R. B., and Cherubini, F. (2015). Quantifying surface albedo and other direct biogeophysical climate forcings of forestry activities. *Global Change Biology*, 21(9), 3246–3266.

Cai, H., Wang, J., Feng, Y., Wang, M., Qin, Z., and Dunn, J. B. (2016). Consideration of land use change-induced surface albedo effects in life-cycle analysis of biofuels. *Energy* and *Environmental Science*, 9(9), 2855–2867.

Caiazzo, F., Malina, R., Staples, M. D., Wolfe, P. J., Yim, S. H., and Barrett, S. R. (2014). Quantifying the climate impacts of albedo changes due to biofuel production: A comparison with biogeochemical effects. *Environmental Research Letters*, 9(2), 024015. doi:10.1088/1748-9326/9/2/024015.

Campbell, G. S., and Norman, J. (2012). *An Introduction to Environmental Biophysics.* Springer Science & Business Media.

Carrer, D., Pique, G., Ferlicoq, M., Ceamanos, X., and Ceschia, E. (2018). What is the potential of cropland albedo management in the fight against global warming? A case study based on the use of cover crops. *Environmental Research Letters*, 13(4), 044030. https://doi.org/10.1088/1748-9326/aab650.

Chen, J., Sciusco, P., Ouyang, Z., Zhang, R., Henebry, G. M., John, R., and Roy, D. P. (2019). Linear downscaling from MODIS to Landsat: Connecting landscape composition with ecosystem functions. *Landscape Ecology*, 34(12), 2917–2934.

Cheng, X., Peng, R., Chen, J., Luo, Y., Zhang, Q., An, S., Chen, J., and Li, B. (2007). CH_4 and N_2O emissions from *Spartina alterniflora* and *Phragmites australis* in experimental mesocosms. *Chemosphere*, 68(3), 420–427.

Cherubini, F., Peters, G. P., Berntsen, T., Strømman, A. H., and Hertwich, E. (2011). CO_2 emissions from biomass combustion for bioenergy: Atmospheric decay and contribution to global warming. *Global Change Biology - Bioenergy*, 3(5), 413–426.

Chu, H., Chen, J., Gottgens, J. F., Ouyang, Z., John, R., Czajkowski, K., and Becker, R. (2014). Net ecosystem methane and carbon dioxide exchanges in a Lake Erie coastal marsh and a nearby cropland. *Journal of Geophysical Research: Biogeosciences*, 119(5), 722–740.

Déry, S. J., and Brown, R. D. (2007). Recent Northern Hemisphere snow cover extent trends and implications for the snow-albedo feedback. *Geophysical Research Letters*, 34(22).https://doi.org/10.1029/2007GL031474.

Forster, P., Ramaswamy, V., Artaxo, P., Berntsen, T., Betts, R., Fahey, D. W., and Nganga, J. (2007). The physical science basis. Contribution of *Working Group I to the Fourth Assessment Report of the Intergovernmental Panel on Climate Change* [Solomon, S., Qin, D. Manning, M., Chen, Z., Marquis, M., Averyt, K. B., Tignor, M., and Miller H. L. (eds.)]. Cambridge University Press, Cambridge, United Kingdom and New York, NY, USA.

Frolking, S., Roulet, N., and Fuglestvedt, J. (2006). How northern peatlands influence the Earth's radiative budget: Sustained methane emission versus sustained carbon sequestration. *Journal of Geophysical Research: Biogeosciences*, 111(G1), doi: 10.1029/2005JG000091.

Gelfand, I., Zenone, T., Jasrotia, P., Chen, J., Hamilton, S. K., and Robertson, G. P. (2011). Carbon debt of Conservation Reserve Program (CRP) grasslands converted to bioenergy production. *Proceedings of the National Academy of Sciences*, 108(33), 13864–13869.

Ghimire, B., Williams, C. A., Masek, J., Gao, F., Wang, Z., Schaaf, C., and He, T. (2014). Global albedo change and radiative cooling from anthropogenic land cover change, 1700 to 2005 based on MODIS, land use harmonization, radiative kernels, and reanalysis. *Geophysical Research Letters*, 41(24), 9087–9096.

Goode, P. R., Qiu, J., Yurchyshyn, V., Hickey, J., Chu, M. C., Kolbe, E., and Koonin, S. E. (2001). Earthshine observations of the Earth's reflectance. *Geophysical Research Letters*, 28(9), 1671–1674.

Hadley, O. L., and Kirchstetter, T. W. (2012). Black-carbon reduction of snow albedo. *Nature Climate Change*, 2(6), 437–440.

Houghton, F. C., and Giddes, R. G. (1992). *U.S. Patent No. 5, 135, 511*. Washington, DC: U.S. Patent and Trademark Office, 200pp.

IPCC (2001). Climate change 2001: synthesis Report. Contribution of *Working Groups I, II, and III to the Third Assessment Report of the Intergovernmental Panel on Climate Change* [Watson, R.T., and the Core Writing Team (eds.)]. Cambridge University Press, Cambridge, United Kingdom, and New York, NY, USA, 398 pp.

IPCC (2007). Climate change 2007: Synthesis report. Contribution of *Working Groups I, II and III to the Fourth Assessment Report of the Intergovernmental Panel on Climate Change* [Core Writing Team, Pachauri, R.K., and Reisinger, A. (eds.)]. IPCC, Geneva, Switzerland, 104 pp.

IPCC (2013). Climate change 2013: The physical science basis. Contribution of *Working Group I to the Fifth Assessment Report of the Intergovernmental Panel on Climate Change* [Stocker, T.F., Qin, D. Plattner, G.-K., Tignor, M., Allen, S.K., Boschung, J., Nauels, A., Xia, Y., Bex V., & Midgley P. M. (eds.)]. Cambridge University Press, Cambridge, United Kingdom and New York, NY, USA, 1535 pp.

Joos, F., Roth, R., Fuglestvedt, J. S., Peters, G. P., Enting, I. G., Bloh, W. V., Brovkin, V., Burke, E. J., Eby, M., Edwards, N. R., Friedrich, T., Frolicher, T. L., Halloran, P. R., Holden, P. B., Jones, C., Kleinen, T., Mackenzie, F. T., Matsumoto, K., Meinshausen, M., Plattner, G.-K., Reisinger, A., Segschneider, J., Shaffer, G., Steinacher, M., Strassmann, K., Tanaka, K., Timmermann, A., and Weaver, A. J. (2013). Carbon dioxide and climate impulse response functions for the computation of greenhouse gas metrics: A multi-model analysis. *Atmospheric Chemistry and Physics*, 13(5), 2793–2825.

Knox, S. H., Jackson, R. B., Poulter, B., McNicol, G., Fluet-Chouinard, E., Zhang, Z., Hugelius, G., Bousquet, P., Canadell, J. G., Saunois, M., Papale, D., Chu, H., Keenan, T. F., Baldocchi, D., Torn, M. S., Mammarella, I., Trotta, C., Aurela, M., Bohrer, G., Campbell, D. I., Cescatti, A., Chamberlain, S., Chen, J., Chen, W., Dengel, S., Desai, A. R., Euskirchen, E., Friborg, T., Gasbarra, D., Goded, I., Goeckede, M., Heimann, M., Helbig, M., Hirano, T., Hollinger, D. Y., Iwata, H., Kang, M., Klatt, J., Krauss, K. W., Kutzbach, L., Lohila, A., Mitra, B., Morin, T. H., Nilsson, M. B., Niu, S., Noormets, A., Oechel, W. C., Peichl, M., Peltola, O., Reba, M. L., Richardson, A. D., Runkle, B. R. K., Ryu, Y., Sachs, T., Schäfer, K. V. R., Schmid, H. P., Shurpali, N., Sonnentag, O., Tang, A. C. I., Ueyama, M., Vargas, R., Vesala, T., Ward, E. J.,

Windham-Myers, L., Wohlfahrt, G., and Zona, D. (2019). FLUXNET-CH_4 synthesis activity: Objectives, observations, and future directions. *Bulletin of the American Meteorological Society*, (2019): 2607–2632.

Lashof, D. A., and Ahuja, D. R. (1990). Relative contributions of greenhouse gas emissions to global warming. *Nature*, 344(6266), 529–531.

Lenton, T. M., and Vaughan, N. E. (2009). The radiative forcing potential of different climate geoengineering options, *Atmospheric Chemistry and Physics*, 9, 5539–5561.

Levasseur, A., Lesage, P., Margni, M., Deschenes, L., and Samson, R. (2010). Considering time in LCA: Dynamic LCA and its application to global warming impact assessments. *Environmental Science and Technology*, 44(8), 3169–3174.

Liu, B. Y., and Jordan, R. C. (1960). The interrelationship and characteristic distribution of direct, diffuse and total solar radiation. *Solar Energy*, 4(3), 1–19.

Manabe, S., and Wetherald, R. T. (1967). Thermal equilibrium of the atmosphere with a given distribution of relative humidity. *Journal of the Atmospheric Sciences*, 24(3), 241–259.

Muñoz, I., Campra, P., and Fernández-Alba, A. R. (2010). Including CO_2-emission equivalence of changes in land surface albedo in life cycle assessment. Methodology and case study on greenhouse agriculture. *The International Journal of Life Cycle Assessment*, 15(7), 672–681.

Myhre, G., Shindell, D., Bréon, F.-M., Collins, W., Fuglestvedt, J., Huang, J., Koch, D., Lamarque, J.-F., Lee, D., Mendoza, B., Nakajima, T., Robock, A., Stephens, G., Takemura T., and Zhang, H. (2013). Anthropogenic and natural radiative forcing supplementary material. In: Climate change 2013: The physical science basis. Contribution of *Working Group I to the Fifth Assessment Report of the Intergovernmental Panel on Climate Change* [Stocker, T.F., Qin, D., Plattner, G.-K. Tignor, M., Allen, S. K., Boschung, J., Nauels, A., Xia, Y., Bex V., & Midgley P. M. (eds.) Available from www.climatechange2013.org and www.ipcc.ch.].

Neubauer, S. C., and Megonigal, J. P. (2015). Moving beyond global warming potentials to quantify the climatic role of ecosystems. *Ecosystems*, 18(6), 1000–1013.

Petrescu, A. M. R., Lohila, A., Tuovinen, J. P., Baldocchi, D. D., Desai, A. R., Roulet, N. T., Vesala, T., Dolman, A. J., Oechel, W. C., Marcolla, B., Friborg, T., Rinne, J., Matthes, J. H., Merbold, L., Meijide, A., Kiely, G., Sottocornola, M., Sachs, T., Zonai, D., Varlagin, A., Lai, D. Y. F., Veenendaal, E., Parmentier, F-J. W., Skiba, U., Lund, M., Hensen, A., van Huissteden, J., Flanagan, L. B., Shurpali, N. J., Grünwald, T., Humphreys, E. R., Jackowicz-Korczynski, M., Aurela, M. A., Laurila, T., Grüning, C., Corradi, C. A. R., Schrier-Uijl, A. P., Christensen, T. R., Tamstorf, M. P., Mastepanov, M., Martikainen, P. J., Verma, S. B., Bernhofer, C., and Cescatti, A. (2015). The uncertain climate footprint of wetlands under human pressure. *Proceedings of the National Academy of Sciences*, 112(15), 4594–4599.

Robertson, G. P., Paul, E. A., and Harwood, R. R. (2000). Greenhouse gases in intensive agriculture: Contributions of individual gases to the radiative forcing of the atmosphere. *Science*, 289(5486), 1922–1925.

Rodhe, H. (1990). A comparison of the contribution of various gases to the greenhouse effect. *Science*, 248(4960), 1217–1219.

Sciusco, P., Chen, J., Abraha, M., Lei, C., Robertson, G. P., Lafortezza, R., Shirkey, G., Ouyang, Z., Zhang, R., and John, R. (2020). Spatiotemporal variations of albedo in managed agricultural landscapes: Inferences to global warming impacts (GWI). *Landscape Ecology*, 35(6), 1385–1402.

Sieber, P., Ericsson, N., and Hansson, P. A. (2019). Climate impact of surface albedo change in life cycle assessment: Implications of site and time dependence. *Environmental Impact Assessment Review*, 77, 191–200.

Shine, K. P., Derwent, R. G., Wuebbles, D. J., and Morcrette, J.-J. (1990). Radiative forcing of climate. In: *Climate Change: The IPCC Scientific Assessment.* [Houghton, J. T., Jenkins, G. J., and Ephraums, J. J.(eds.)] Cambridge University Press, Cambridge, United Kingdom and New York, NY, USA, 41–68.

Shine, K. P., Fuglestvedt, J. S., Hailemariam, K., and Stuber, N. (2005). Alternatives to the global warming potential for comparing climate impacts of emissions of greenhouse gases. *Climatic Change*, 68(3), 281–302.

Skytt, T., Nielsen, S. N., and Jonsson, B. G. (2020). Global warming potential and absolute global temperature change potential from carbon dioxide and methane fluxes as indicators of regional sustainability—a case study of Jämtland, Sweden. *Ecological Indicators*, 110, 105831.

Syakila, A., and Kroeze, C. (2011). The global nitrous oxide budget revisited. *Greenhouse Gas Measurement and Management*, 1(1), 17–26.

Sundquist, E. T. (1993). The global carbon dioxide budget. *Science*, 259(5097), 934–941.

Smith, S. J., and Wigley, M. L. (2000). Global warming potentials: 1. Climatic implications of emissions reductions. *Climatic Change*, 44(4), 445–457.

Trenberth, K. E., Fasullo, J. T., and Kiehl, J. (2009). Earth's global energy budget. *Bulletin of the American Meteorological Society*, 90(3), 311–324.

Zenone, T., Chen, J., Deal, M. W., Wilske, B., Jasrotia, P., Xu, J., Bhardwaj, A., Hamilton, S., and Robertson, G. P.(2011). CO_2 fluxes of transitional bioenergy crops: Effect of land conversion during the first year of cultivation. *Global Change Biology - Bioenergy*, 3(5), 401–412.

Index

H

I

J

K

L

M

N

P

Q

R

S

Author

Jiquan Chen
Landscape Ecology & Ecosystem Science (LEES) Lab
Department of Geography, Environment, and Spatial Sciences &
Center for Global Change and Earth Observations
Michigan State University
East Lansing, MI 48823
Email: jqchen@msu.edu
Web: http://lees.geo.msu.edu

Contributing Authors

Ge Sun
Eastern Forest Environmental Threat Assessment Center
USDA Forest Service
Email: ge.sun@usda.gov

Cheyenne Lei
Landscape Ecology & Ecosystem Science (LEES) Lab
Department of Geography, Environment, and Spatial Sciences &
Center for Global Change and Earth Observations
Michigan State University
East Lansing, MI 48823
Email: cheyenne@msu.edu

Pietro Sciusco
Landscape Ecology & Ecosystem Science (LEES) Lab
Department of Geography, Environment, and Spatial Sciences
Michigan State University
East Lansing, MI 48823
Email: sciuscop@msu.edu